A Basic Guide to RCRA

Understanding Solid and Hazardous Waste Management

Second Edition

Salvatore Caccavale, CPEA

AMERICAN SOCIETY OF
SAFETY PROFESSIONALS

Print ISBN: 978-1-885581-76-1

Printed in the United States

30 29 28 27 26 25 24 6 7 8 9 10 11 12

Library of Congress Cataloging-in-Publication Data
Caccavale, Salvatore.
A basic guide to RCRA : understanding solid and hazardous waste management / Salvatore Caccavale, CPEA. -- 2nd ed.
p. cm.
Includes bibliographical references and index.
ISBN 978-1-885581-76-1 (alk. paper)
1. Hazardous wastes--United States--Management. 2. United States. Resource Conservation and Recovery Act of 1976. I. Title.
TD1040.C33 2013
363.72'87560973--dc23
2013013330

Dedication

This book is dedicated to
Danielle, Tabetha, Gavin, Zachary, Elliott, Ashley,
Riley and Oliver
and
in loving memory of
Angelo F. and Mary Caccavale.

About the Author

Salvatore Caccavale, CPEA, has been employed in various corporate environmental, health, safety, and security leadership roles for Fortune 500 companies such as BASF Corporation, Valspar Corporation, Rexam Beverage Can Americas, Superior Bulk Logistics, Inc., A. M. Castle and Co., Tradebe Environmental Services, Kellogg's, and First Group America for more than twenty years.

Mr. Caccavale is a Certified Professional Environmental, Health, and Safety Auditor and a former master's level Certified Hazardous Materials Manager. Mr. Caccavale graduated from Monmouth University in West Long Branch, New Jersey, with a BS in Business Management. He has served in several executive office positions with the National Safety Council, most notably as vice president of the Business and Industry Division and two terms on the board of directors.

Mr. Caccavale has instructed EHS classes for both the Community College of Beaver County in Center Township, Pennsylvania, and the University of Medicine and Dentistry in Piscataway, New Jersey. He is a noted speaker and guest columnist for professional publications. Mr. Caccavale serves on the editorial review boards of the American Society of Safety Professionals' *Professional Safety* journal and the National Safety Council's *Safety+Health*.

Contents

PREFACE

I can imagine that most professionals in the environmental field will look at this book and think, "Oh, no, not another book on RCRA!" However, let me state that I wrote this book by living through the trials and errors of implementing and maintaining an RCRA program.

As more and more companies in the new millennium downsize, reengineer, or restructure, depending on your choice of buzzword, safety and health professionals, who have had little to no understanding of the environmental disciplines, specifically RCRA, are now being assigned to positions in which they have full responsibility for the hazardous and solid waste program for either the facility at which they work or their entire corporation.

This book is dedicated to those individuals by an individual who experienced that same dilemma firsthand. I hope the real-life scenarios in this book, in conjunction with the regulatory verbiage, will help the novice-to-intermediate professional obtain a better understanding of how the RCRA regulations work and provide some helpful tips on how to stay out of regulatory hot water.

Safety professionals have the benefit of dealing with performance-oriented OSHA standards; the regulatory agency provides the skeleton of a regulation and allows the industry and/or company to decide how the regulation applies.

In the environmental field, however, I have found that the same is not always true with regard to regulations. First, there are copious numbers of acronyms to understand and piece together, and second, all of the restrictions are based on technical data (e.g., permit limits, laboratory results). RCRA is no different; there are statutory and regulatory stipulations to comprehend, and if they are not daunting enough, throw in a few state restrictions.

This book moves through the federal RCRA regulation using a systematic approach, from an explanation of the regulation to how it applies to an operating facility.

Chapter 1

The Environmental Evolution Legislative History

Chapter 1 Learning Objectives

- Understand how and why the Resource Conservation and Recovery Act (RCRA) came about.
- List the main RCRA regulation subtitles and their formal applications.
- List the goals of the federal RCRA regulation program.

With the advent of the industrial revolution in the late 1800s and early 1900s, our world was treated to many wonders that we now, in the twenty-first century, take for granted. Clothing, transportation, power tools, sports equipment, and cellular phones—the list of products that have made our lives easier goes on and on.

The flip side to these miraculous achievements is the generation of waste. Today, consumer product companies are trying to minimize waste by offering many types of reusable packaging. Recycling has taken on a larger role in society, with many states and communities implementing mandatory programs to reduce the amount of solid waste consumers place in landfills.

Back in the early days of the industrial revolution, there were no Treatment, Storage, and Disposal Facilities (TSDFs) to manage and/or destroy a company's hazardous waste. Companies weren't even required to determine whether the waste they were generating at the end of the production line was categorized as hazardous. The solution for disposing of most of the waste was to dig a big hole somewhere on the property and fill it with the waste generated by production. When that hole was filled, cap it off and dig another hole. The saving grace for society was that some companies had the environmental insight to apply liners, either natural (clay) or man-made (plastic or cloth), where they were dumping their waste to at least attempt to catch the leachate.

For years, industries claimed they didn't realize what they were putting into the ground would have any effect on the environment. Since the advent of the environmental disciplines (mid-1960s), our society now realizes what a mistake that approach was. Even today, abandoned industrial sites and misused landfills are still being cleaned up through the Environmental Protection Agency's (EPA) Superfund program. Industry was not the only culprit, however; mom-and-pop

Exhibit 1. **How the U.S. Government Tracks the Life Cycles of Chemicals**

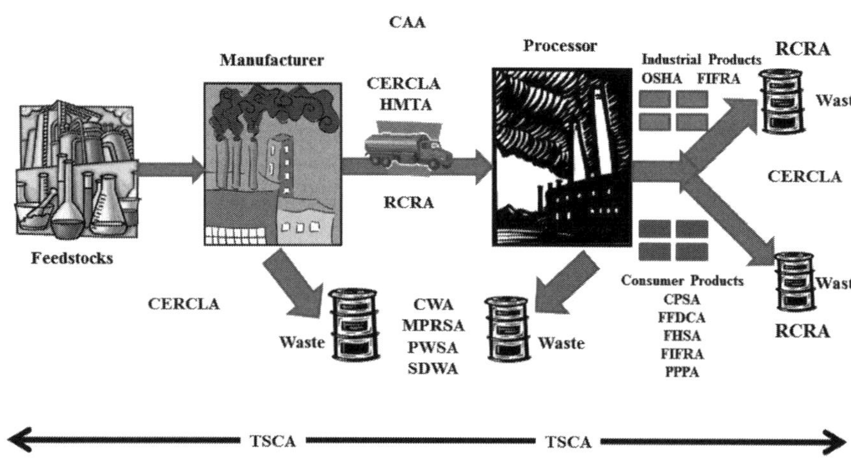

HMTA	Hazardous Materials Transportation Act
CAA	Clean Air Act
MPRSA	Marine Protection Research & Sanctuaries Act
CPSA	Consumer Product Safety Act
OSHA	Occupational Safety and Health Act
CWA	Clean Water Act
PPPA	Poison Prevention Packaging Act
FFDCA	Federal Food, Drug and Cosmetic Act
PWSA	Ports and Waterways Safety Act
FHSA	Federal Hazardous Substance Act
RCRA	Resource Conservation and Recovery Act
FIFRA	Federal Insecticide, Fungicide and Rodenticide Act
SDWA	Safe Drinking Water Act
CERCLA	Comprehensive Environmental Response, Compensation and Liability Act
TSCA	Toxic Substance Control Act

shops, farmers, dry-cleaning operations, and other types of small businesses had just as much effect on polluting the environment.

The Advent of RCRA

In the early 1970s, the story of Love Canal, a housing development built over a capped industrial dump site in suburban Niagara Falls, New York, ran rampant in the media and took the public by surprise. Concern that this type of event could happen in "our backyard" led to public demand for legislative bodies to develop regulations that would ensure the proper management of industrial wastes (Exhibit 1).

In 1965, Congress passed the Solid Waste Disposal Act (SWDA) (42 U.S.C. 6912) for the purpose of improving solid waste disposal methods. SWDA was amended several times, most significantly by the Resource Conservation and Recovery Act (RCRA) of 1976 (42 U.S.C. 6921) and the Hazardous and Solid Waste Amendments (HSWA) of 1984; these acts collectively are known as RCRA.

RCRA was designed to establish a national program to protect the natural resources of the United States from the improper handling and storage of hazardous wastes. Congress gave the task of tracking this regulation to the EPA. The codified RCRA regulations can be found in Title 40 of the Code of Federal Regulations (CFR), Parts 240 through 282.

The main components of RCRA include the identification of hazardous waste; manifest tracking of waste "cradle-to-grave"; operating standards for generators, transporters, and TSDFs; a permit system for TSDFs; and authorization of to assist in implementing the program.

The current goals set forth by RCRA are

- to protect human health and the environment from the potential hazards of waste disposal;
- to conserve energy and natural resources;
- to reduce the amount of hazardous waste generated;
- to ensure that wastes are managed in an environmentally sound manner;
- to prevent future problems caused by irresponsible waste management;
- to clean up releases of hazardous waste in a timely, flexible, and protective manner.

To achieve these goals, three distinct yet interrelated sections (called subtitles) exist under RCRA. (Other subtitles exist in the RCRA regulation, but these three have the most impact, and they will be discussed in this book.)

- **Subtitle C**—Establishes a system for controlling hazardous waste from the time of generation until ultimate disposal.
- **Subtitle D**—establishes a system for controlling solid (primarily non-hazardous) waste, such as household waste.

- **Subtitle I**—Established by HSWA, regulates toxic substances and petroleum products stored in underground tanks.

One other subtitle and one subpart are worth mentioning:

- **Subtitle J**—In 1988, the EPA attempted to enact the tracking and management of medical waste under the RCRA regulation but following a two-year demonstration program decided not to pursue the issue. At present no federal EPA regulations are in effect for medical waste; however, many states have adopted their own programs.

Land Disposal Restriction (LDR) (40 CFR 268)

One of the major impacts HSWA had on the implementation of the RCRA program was the restriction of land disposal for certain wastes. These land disposal restrictions are commonly referred to as *land ban*, and the hazardous wastes affected are called *restricted wastes*. The EPA banned land disposal for certain hazardous wastes that can migrate through soil and pollute groundwater. Hazardous wastes covered by the land ban include liquid metals, free cyanides, dioxin-containing wastes, and discarded chemical products such as xylene, formic acid, and methyl alcohol. The ban also prohibits land disposal of diesel fuel, hydrochloric acid, and used solvents without proper treatment or certification of limited recycling operations.

Although RCRA creates a framework for the proper management of hazardous and nonhazardous solid waste, it does not address the problems of hazardous waste encountered at inactive or abandoned sites or those resulting from spills that require an emergency response. These issues are addressed by the Comprehensive Environmental Response, Compensation and Liability Act (CERCLA) (40 CFR Part 280.12), otherwise known as Superfund. We will further discuss the interaction between RCRA and CERCLA later in the book.

Subtitle C (Hazardous Waste Management)

Subtitle C establishes a comprehensive cradle-to-grave program to regulate hazardous waste from generation through proper disposal or destruction. Generators of hazardous waste are the first link in the cradle-to-grave chain of hazardous waste. Generators of more than 100 kilograms (220 pounds) of hazardous waste, or 1 kilogram (2.2 pounds) of acutely hazardous waste a month must comply with all generator requirements under Subtitle C (40 CFR Part 262).

Subtitle D (State or Regional Solid Waste Plans)

The SWDA established grant programs for the development of solid waste plans by states and/or interstate agencies. Subsequent amendments to the SWDA have substantially increased the EPA's involvement in solid waste management. However, the Subtitle D (40 CFR Part 255) program continues to be implemented by state and local governments.

The term *solid waste* used in Subtitle D refers almost exclusively to non-hazardous solid waste. The primary objectives of Subtitle D are to encourage environmentally sound solid waste management practices, maximize the reuse of valuable recoverable resources, and foster resource conservation.

Subtitle I (Regulation of Underground Storage Tanks)

Congress enacted Subtitle I (40 CFR Part 280) to control and prevent leaks from underground storage tanks (USTs). The UST program broke new ground for RCRA, as for the first time the RCRA program applied to products as well as wastes. Specifically, Subtitle I regulates underground tanks storing regulated substances, including petroleum products (e.g., gasoline and crude oil) and Superfund-defined hazardous substances. Tanks storing hazardous waste, however, are regulated under Subtitle C.

Summary

Who must comply with RCRA? Any organization that generates, transports, treats, stores, or disposes of hazardous waste. Any organization that produces, burns, distributes, or markets any hazardous waste-derived fuels must notify the EPA of their activities and comply with RCRA. Any organization that stores hazardous materials, as defined in CERCLA (Superfund), in underground tanks must also notify EPA and comply with RCRA.

From its inception, RCRA has been able to control the flow of hazardous wastes generated throughout the United States. Our society mismanaged both hazardous and nonhazardous wastes for many years. Today we focus on programs like waste minimization and pollution prevention, where both industry and communities are discovering new ventures to help reduce and recycle wastes.

Most of the remaining chapters in this book will focus on Subtitle C and the importance of ensuring that the generator requirements of the RCRA regulations are met.

Chapter 2

Applicable Responsibilities

Chapter 2 Learning Objectives

- Distinguish between the regulations governing hazardous waste generators, transporters, and TSDFs.
- Understand the federal hazardous waste classification system for generators.
- Understand the permit application and management system requirements for TSDFs.
- Understand the regulatory enforcement options under the RCRA regulation for generators, transporters, and TSDFs.
- Recognize common mistakes noted by regulatory inspectors that are considered red flags.

The three main players under the RCRA umbrella are hazardous waste generators, transporters, and TSDFs. In this section we will discuss each of the parties' roles and responsibilities under the RCRA regulation.

Generator

The level of regulatory burden placed on a hazardous waste generator is determined by the amount of hazardous waste it generates monthly. There are three categories of generators:

- Conditionally Exempt Small Quantity Generators (CESQGs) (40 CFR Part 261.5)
- Small Quantity Generators (SQGs) (40 CFR 262.34 (d))
- Fully Regulated or Large Quantity Generators (LQGs) (40 CFR 262.34(a))

A generator is classified CESQG if it generates less than or equal to 100 kilograms (220 pounds) of hazardous waste per month or less than or equal to 1 kilogram (2.2 pounds) of acutely hazardous waste per month and no more than 1000 kilograms (2200 pounds) of hazardous waste has been accumulated.

NOTE: Acute hazardous wastes listed in 40 CFR 261.30 are those wastes that, in small doses, are capable of causing death or significantly contributing to irreversible and/or incapacitating illness (e.g., phosgene).

Regulatory relief for CESQGs includes no EPA identification (ID) number required, no waste manifest necessary during shipments, and no land disposal restrictions.

A generator classified as SQG is one that generates more than 100 kilograms but less than 1000 kilograms of hazardous waste in a month, has accumulated no more than 6000 kilograms (13,200 pounds) of hazardous waste, and never exceeds 1 kilogram (2.2 pounds) of an acute hazardous waste.

Regulatory relief for SQGs includes no written contingency and training plans required, may accumulate hazardous waste for up to 180 days (270 days if waste must be shipped more than 200 miles), and no air emission control requirements under 40 CFR 265 Subpart CC (discussed in Chapter 4).

An LQG generates 1000 kilograms or more per month of hazardous waste. Allowances for an LQG includes no limit on the amount of waste that can be accumulated on-site; an LQG can accumulate on-site up to 90 days without a permit or interim status. Requirements for an LQG include a written contingency plan, a written training program, and the control of air emissions per 40 CFR 265 Subpart CC.

A generator's status and specific requirements may change from month to month depending on chemicals used and subsequent waste generated. This is referred to as *episodic generation*. Rather than contending with the regulatory definition of SQG or LQG, most facilities choose to take the conservative approach and file as an LQG. The benefits of being compliant with the LQG regulations outweigh the benefits of juggling SQG status each month and possibly being out of compliance when an inspector shows up.

If a generator whose status is SQG has a one-time event (a chemical inventory cleanup or a load of off-spec material, for example) where waste shipment would exceed SQG limits, it can petition the EPA or state agency for an exemption. Such exemptions are handled on a case-by-case basis.

For all generators of hazardous waste, the following requirements (40 CFR Part 262) must be established as part of the company's or facility's RCRA program:

- record-keeping practices that accurately identify the quantities of hazardous waste generated, the composition of the wastes that are significant in quantity or in potential harm to human health or the environment, and the disposition of such waste

- labeling practices for containers used in the storage, transportation, or disposal of hazardous waste that accurately identify the material inside
- use of appropriate containers for hazardous wastes
- provision of accurate information about the general chemical composition of hazardous waste to persons transporting, treating, storing, or disposing of such wastes
- use of a manifest system and any other reasonable means necessary to assure that all hazardous waste generated is designated for treatment, storage, or disposal in, and arrives at, the TSDF for which the permit has been issued
- submission of reports to the administrator at least every two years describing
  - (A) the quantity and nature of hazardous waste generated during the year;
  - (B) the disposition of all hazardous wastes reported under A;
  - (C) the efforts taken during the year to reduce the volume and toxicity of waste generated; and
  - (D) the changes in volumes and toxicity of waste actually achieved during the reporting year in question compared with previous years.

A generator of hazardous waste has the responsibility to determine where the company/facility fits into the regulation based on the amount of hazardous waste generated. Determining whether or not a waste is hazardous will be discussed in the next chapter. A solid pollution prevention program helps to minimize both environmental liability to a company and costs associated with disposal of hazardous wastes.

Under 40 CFR 262.12, a generator of hazardous waste must not treat, store, dispose of, transport, or offer the waste for transportation without having received an EPA identification number from the administration. A facility must apply using EPA Form 8700-12 for assignment of an identification number. (See the example of a completed form and instructions in Appendix V.) Information on who must complete the form and how to complete it are also in the appendix. Additional information regarding EPA ID numbers, including forms and instructions, can be found at the following URL: http://www.cpa.gov/epawaste/index.htm.

The generator must not offer hazardous waste to a transporter or TSDF that has not received an EPA ID number, so it is important that the generator obtain an EPA ID number prior to startup of a process, laboratory, or the like, and ensure that the transporter and TSDF both have EPA ID numbers prior to offering your hazardous waste for shipment.

Transporter

A transporter of hazardous waste must comply with both the transporter's regulations (40 CFR 263) under the RCRA hazardous waste program (delivery, cleanup during transit, spill reporting) and the licensing procedures for each state through which it will transport the hazardous waste. The transporter must accept waste only in accordance with the manifest system and comply with all DOT regulations as they apply to hazardous waste shipments (such as driver qualifications, hazmat employee training, and insurance). If the transporter stores waste at a transfer facility, the maximum amount of storage time is ten days. The transporter must retain its copy of the hazardous waste manifest for a minimum of three years.

TSDFs

Any facility that treats, stores, or disposes of hazardous waste is considered a TSDF. Requirements have been established under the RCRA regulation for TSDFs (40 CFR 264, 265 and 270).

A TSDF must obtain approval to operate from the U.S. EPA or authorized state agency. There is a two-step approval process before a facility can be classified as a permitted TSDF. The Part A or interim status permit requires a facility to obtain an EPA ID number and submit summary information to the agency(s), which includes design, construction, operation, and maintenance of the facility. A TSDF may operate under a Part A permit until the Part B application is either approved or denied.

The Part B permit allows a TSDF to accept, store, treat, and dispose of hazardous waste with the following requirements:

- a waste analysis plan for incoming shipments
- security measures
- a documented inspection program for malfunctions, operator errors, and discharges
- personnel training
- for ignitable, reactive, and incompatible wastes, separate storage areas and handling, cleaning and filling of containers specified in §264.176 and §264.177
- preparedness for and prevention of emergencies and releases
- a written contingency plan and emergency procedures
- written operating records, manifest records, and biennial reports of facility activities
- groundwater protection for land disposal facilities
- closure and post-closure plans
- financial information regarding closure, post-closure, accidents, and bankruptcy.

TSDFs must be built in locations away from natural disasters such as floods, hurricanes, and earthquakes. The facility must also be able to accept wastes in accordance with manifest rules; manage containers in accordance with container rules; manage tanks in accordance with tank rules; manage other units (surface impoundments, landfills, incinerators, and so on) in accordance with applicable technical standards; control, monitor, and document volatile organic air emissions (fugitive emissions); comply with land disposal restrictions (where applicable); and maintain adequate insurance during operation and adequate funding for closure and post-closure.

Some TSDF companies provide their own transportation services; most subcontract with licensed hazardous waste haulers. Even though a hazardous shipment has left a company's premises under the cradle-to-grave concept, the company has liability for the shipment until the waste is officially disposed of. Companies using transporters and TSDFs must audit or at least visit the transporter's terminal and TSDF (including cement kilns) or hire a third-party auditor to conduct the process prior to sending waste off-site. It is a worthwhile investment.

A company or facility does not have to apply for a RCRA permit if it falls under one of the following specific exemptions:

- It is a generator that accumulates hazardous waste on-site for less than 90 days.
- It generates less than 100 kilograms (220 pounds) of hazardous waste in a calendar month and treats, stores or disposes of it in compliance with all RCRA hazardous waste requirements.
- It owns or operates a totally enclosed treatment facility—for example, a hazardous waste treatment facility that is directly connected to an industrial production process and is constructed and operated to prevent the release of hazardous waste into the environment during treatment (e.g., a pipe in which waste acid is neutralized).
- It is a transporter that stores manifested shipments of hazardous waste for ten days or less in permitted containers at a transfer facility.

The aforementioned are the federal RCRA requirements; check with your state for more stringent requirements.

Regulatory Enforcement

The EPA or authorized state has three enforcement options under RCRA:

- Administrative sanctions or penalties
- Civil penalties
- Criminal penalties

Administrative Sanctions or Penalties—These are nonjudicial enforcement actions taken by either the EPA or the state. There are two types: *informal actions* and *administrative orders* (i.e., compliance orders, corrective action orders). The maximum penalty is $25,000 per day of noncompliance for each violation.

Civil Penalties –These are formal lawsuits against individuals and companies who have failed to comply with some statutory/regulatory requirement or administrative order. There are four types of civil penalties: compliance action; corrective action; monitoring and analysis; and imminent hazard. The maximum penalty is the same as for administrative sanctions: $25,000 per day of noncompliance for each violation.

Criminal Penalties –Seven criminal acts under RCRA carry severe criminal penalties. These acts are knowingly

- transporting waste to a non-permitted facility;
- treating, storing, or disposing of waste without a permit;
- omitting information or making a false statement on any application, label, manifest, record, report, permit, or compliance document;
- not complying with record-keeping and reporting requirements;
- transporting without a manifest;
- exporting waste without the consent of the receiving country; and
- taking action resulting in imminent danger.

The maximum penalties are:

- $50,000 per day of violation and two years in prison
- $100,000 per day and four years in prison (repeat offenses)
- $250,000 per day and fifteen years in prison for an individual or $1 million for an organization (applicable to "taking action resulting in imminent danger")

Repeat offenses may include instances of offenses throughout a company even if the company has facilties in different locations. For example, if inspectors found a problem at company's New York facility and later found the same type of problem at its Texas facility, they could consider the Texas infraction a repeat offense, with a higher fine and longer prison sentence. The moral is that if a violation is discovered at one facility, a mechanism should be in place to communicate information about it to all facilities within the company.

Common Mistakes (Things Inspectors Can Trip Over)

- not covering/closing containers
- incorrect container labeling
- improper use of regulated chemicals
- failure to record accumulation start date
- exceeding the three-day time limit for full satellite-facility containers
- lack of adequate secondary containment
- improper training documentation (e.g., job descriptions, titles/qualifications, frequency)
- failure to conduct inspections
- failure to update contingency plan

Summary

In this chapter the three participants under the RCRA regulation—generators, transporters, and TSDFs—were described. All three have requirements set forth by the regulation regarding their roles in the cradle-to-grave process of hazardous waste. Generators are classified by the amount of hazardous waste they generate in a calendar month, and different levels of requirements are based on those classifications. The basics of the transporter were described, and complicated permitting requirements were set forth for anyone or any company that chooses to accept, store, treat, and/or dispose of hazardous waste. Finally, there was a discussion of regulatory enforcement and how fines gradually increase based on the severity of the infraction.

Chapter 3

Waste Determination

Chapter 3 Learning Objectives

- Be able to determine whether a waste stream is hazardous.
- Recognize the difference between *listed* and *characterized* waste streams.
- Understand how *mixture, derived-from*, and *contained-in* rules apply to waste streams.
- Understand what the land disposal restriction (LDR) requirements are and know how they are applied.
- Determine how to manage used oil and universal waste streams.

Before attempting to determine whether waste is hazardous, consult the EPA guidelines to determine whether the waste is a *solid* waste. Though most solid waste management programs are under the guidance of individual state programs, the EPA set forth in 40 CFR 240 and 243 some mandatory and advisory guidelines regarding the implementation of a solid waste management program.

For example, thermal processing facilities (incinerators) processing 50 tons or more of solid waste per day fall under 40 CFR 240 and must meet minimum requirements for waste acceptance; facility design and operations; and discharges into streams, rivers and the atmosphere. The EPA also insists that thermal process residue (ash) be disposed of in a responsible manner.

Storage and collection guidelines (40 CFR 243) apply to residential, commercial, and institutional solid and street wastes, while source separation for material recovery guidelines (40 CFR 246) apply to residential, commercial, and institutional wastes. Individual state plan guidance is promulgated in 40 CFR 256.

Waste Identification/Classification

RCRA provides both a regulatory and a statutory definition of hazardous waste. Hazardous waste is a subset of solid waste; therefore, a material cannot be classified as hazardous waste if it is not classified as solid waste (Exhibit 2). RCRA defines solid waste as

> "Any garbage, refuse, sludge from a wastewater treatment plant or pollution control facility, and other discarded material, including solid, liquid, semisolid, or contained gaseous material, resulting from industrial, commercial, mining, and agricultural operations and from community activities."

Exhibit 2. **Solid and Hazardous Waste Universe**

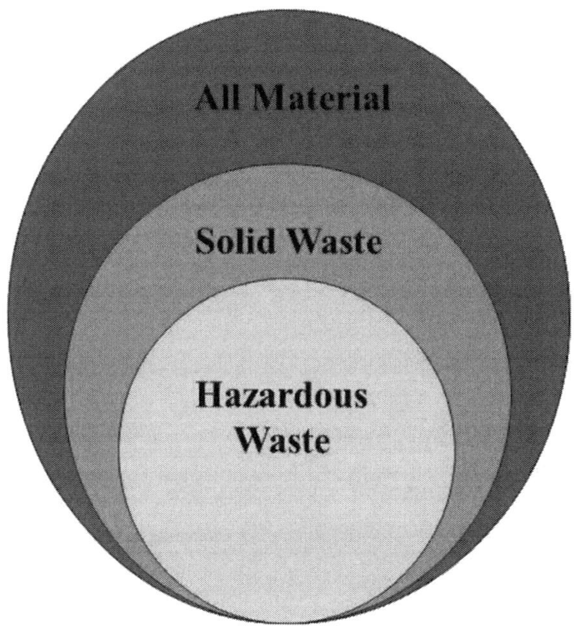

There are four other ways in which materials can be classified as solid waste:

- If they are destined for disposal (40 CFR 261.2(b)):

  If a material is disposed of, burned or incinerated, treated (but not recycled), or stored before being disposed of or burned or incinerated, then the material is considered a waste.

- If they are recycled in certain ways (40 CFR 261.2(c)):

  If a material is being recycled in a manner specified in 40 CFR 261.2(c), then it is considered a waste. Table 1 in 40 CFR 261.2 summarizes the types of recycling that make a material a waste and those that do not.

- If they are inherently waste like materials (40 CFR 261.2(d)):
  The EPA lists a number of materials that are considered to be inherently waste like materials. This means these materials are automatically wastes unless excluded, and include specific dioxins being discarded.
- If they are military munitions (40 CFR 261.2(a)(4)):
  All ammunition products and components produced for or used by the United States Department of Defense for national defense or security are considered solid waste.

Municipal solid waste is a subset of solid waste defined as durable goods (such as appliances, tires, and batteries), nondurable goods (such as newspapers, books, and magazines), containers and packaging, food wastes, yard trimmings, and miscellaneous organic wastes from residential, commercial, and industrial nonprocess sources (see example of Solid Waste Recycling Program in Appendix VI).

Industrial waste is also a subset of solid waste and is defined as solid waste generated by manufacturing or industrial processes that is not regulated under Subtitle C of RCRA. Such waste may include, but is not limited to waste resulting from

- generating electric power;
- manufacture of fertilizer or agricultural chemicals; food and related products or by-products; inorganic chemicals; iron and steel; leather and leather products; nonferrous metals; organic chemicals; plastics and resins; pulp and paper; rubber and miscellaneous plastic products; stone, glass, clay, and concrete products; textiles; transportation equipment; and
- treating water.

Solid wastes are generally defined as hazardous wastes in one of two ways (40 CFR 261). First, solid wastes are considered hazardous wastes if they are listed as such by the EPA in 40 CFR 261 Subpart D. Second, solid wastes are considered hazardous wastes if they exhibit any of the four characteristics found in 40 CFR 261 Subpart C. These characteristics are ignitability, corrosivity, reactivity, and toxicity (based on the Toxicity Characteristic Leaching Procedure (TCLP—EPA Test Method 1311).

The EPA has determined (40 CFR 261.2(e)) that a material is not considered a solid waste if it is reused in one of three ways:

- as an ingredient to make a product
- as an effective substitute for a product or
- returned to the process that generated it without first being reclaimed or land-disposed

Now that we have defined a *solid* waste, let's define a *hazardous* waste. The RCRA statutory definition of a hazardous waste is as follows:

"A solid waste or combination of solid wastes, which because of its quantity, concentration, or physical, chemical, or infectious characteristics may:

1. cause, or significantly contribute to an increase in mortality or an increase in serious irreversible, or incapacitating reversible, illness; or
2. pose a substantial present or potential hazard to human health or the environment when improperly treated, stored, transported, or disposed of, or otherwise managed."

Hazardous wastes can be both *listed* and *characteristic.*

Listed Wastes

Listed wastes are wastes from generic industrial processes, wastes from certain sectors of industry, and unused pure chemical products or formulations. Because these wastes are dangerous enough to warrant full Subtitle C regulation based on their origin, any waste fitting a narrative listing description is considered a listed hazardous waste.

A solid waste is hazardous if it is named on one of three lists promulgated by the EPA. The EPA developed these lists by examining different types of wastes and chemical products to see if they met any of the following criteria:

- They exhibit one of the four characteristics (ignitable, toxic, corrosive, reactive) of a hazardous waste.
- They contain such dangerous chemicals that they pose a threat to human health and the environment even when properly managed. These wastes are fatal to humans and animals. Such wastes are known as *acute hazardous waste.*
- They contain toxic chemicals at levels that could pose a threat to human health and the environment even if properly managed. These wastes are known as *toxic listed wastes.*

The EPA hazardous waste lists are divided into three groups:

- *Generic process wastes* are solid wastes from nonspecific sources that are listed as hazardous wastes in 40 CFR 261.31. These wastes are assigned "*F*" waste codes. Examples from the *F* list include spent halogenated solvents used in degreasing, such as carbon tetrachloride (F001); paint strippers such as methylene chloride (F002); dry-cleaning filters, such as tetrachloroethylene (F002); and nonhalogenated solvents, such as xylene or acetone (F003), cresols (F004), and toluene, carbon disulfide, and benzene (F005).
- *Industry-specific wastes* are solid wastes from specific sources that are listed as hazardous wastes in 40 CFR 261.32. Industry-specific wastes are assigned K waste codes. Examples of this type of waste are sludge, still bottoms, spent catalysts, and residues from industry, such as organic chemicals (K009, K017) and explosives (K044–K047).

- *Commercial chemical product wastes* are listed wastes in 40 CFR 261.33. This list describes wastes that are discarded formulations in which the listed constituent is the sole active ingredient. Commercial chemical product wastes are assigned *P* or *U* waste codes. Examples of *P* and *U* listed wastes include chemicals such as acetone (U002), chloroform (U044), creosote (U051), phosgene (P095), sodium cyanide (P106), and certain pesticides, such as DDT (U061).

Characteristic Wastes

Characteristic wastes exhibit measurable properties that indicate that they pose enough threat to warrant regulation as a hazardous waste. The EPA identified characteristics that, when present in a waste, can cause death or injuries to humans or lead to ecological damage. The characteristics identify both acute (short-term) and chronic (long-term) hazards and are an essential supplement to the hazardous waste listings.

There are four generic properties or characteristics that can cause a waste to be regulated as hazardous waste. Each has a specific regulatory definition (40 CFR 261 Subpart C):

Ignitability (D001) –a waste that can burn

- Liquids with a flash point of less than 140°F
- Nonliquid materials with the potential to spontaneously combust
- DOT ignitable compressed gases
- DOT oxidizers

Corrosivity (D002) –an acid or base that can corrode or dissolve flesh, metal, or other materials

- Aqueous solutions with pH less than or equal to 2 or greater than or equal to 12.5
- Liquids that corrode steel at the rate of more than .25 inch per year.

Reactivity (D003) –a waste that can explode when mixed with air, water, or other chemicals

- Normally unstable and undergoes violent change without detonating
- Reacts violently with water
- Forms potential explosive mixtures with water
- Produces toxic gases when mixed with water
- Cyanide- or sulfide-containing material that releases toxic gases under a specified pH range
- Capable of detonation if struck or heated
- Capable of detonation under standard temperature and pressure
- A DOT explosive, as specified

Toxicity (D004 – D043) – a waste that is harmful to living organisms

Includes any waste that, when subject to the TCLP, produces a leachate that contains one or more of 40 constituents (organic, metal, or pesticide-bearing wastes) listed in Table 1 of 40 CFR 261.24 at concentrations equal to or greater than those specified in the table.

The TCLP is designed to simulate landfill conditions. If a waste could contaminate groundwater with one or more of 40 constituents if placed in a landfill, then it exhibits the toxicity characteristic for each constituent that could leach out. Some of the contaminants included on this list are arsenic (D004), benzene (D018), cresol (D026), lead (D008), mercury (D009), pyridine (D038), silver (D011) and vinyl chloride (D043).

Let's take a look at an example of how a material waste, methyl ethyl ketone (MEK) (D035 < 0.00 milligrams per liter), can be either a listed or characteristic hazardous waste, depending on how it is used:

Process Generating Waste	EPA Code
Paint that contains MEK	D001, D035
Contaminated raw material	U159, D001, D035
Solvent mixture used in process clean-up	F005, D001, D035

"Mixture" Rule (40 CFR 261.3)

One of the questions the EPA faced when setting conditions for identifying hazardous wastes was how to classify a waste mixture that contained both a listed hazardous waste and a nonhazardous solid waste. The EPA decided that any waste mixture containing a listed hazardous waste must be considered a hazard-

Exhibit 3. **"Mixture" Rule" Example Involving Two Liquids**

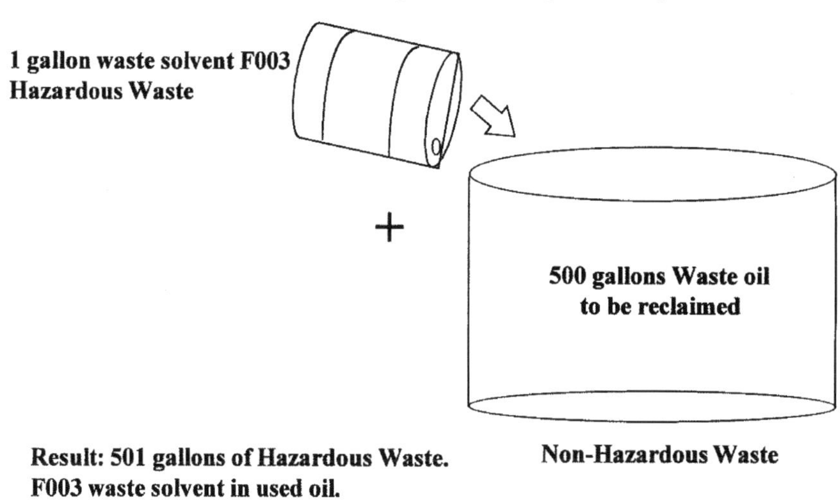

ous waste and must be managed accordingly. This applies regardless of what percentage of the mixture is composed of a listed hazardous waste (Exhibit 3).

However, if you mix a characteristic hazardous waste with a nonhazardous solid waste, testing must be conducted to determine whether the new waste combination is considered a hazardous waste (Exhibit 4).

"Derived-From" Rule (40 CFR 261.3 (c)(d))

Another issue that confronted the EPA was how to classify residues from the treatment, storage, or disposal of a listed hazardous waste. The EPA decided that any residue "derived from" a listed hazardous waste must be considered a hazardous waste. The EPA established this rule because it is a reasonable assumption that wastes derived from hazardous wastes are themselves hazardous unless it is proven that they will not pose a threat to human health and the environment (Exhibit 5).

Let's look at examples of the Mixture and Derived-From rules again using MEK.

Mixture Rule:

Example: Paint containing MEK and used oil (D035, D001). You have to test to determine if it is a hazardous a waste stream or not.

Derived-From Rule:

Example: Spent MEK (F005) solution reclaimed in distillation unit. If reusable MEK is used in product, then there is no hazardous waste; but any still bottoms are hazardous wastes (F005, D035).

Exhibit 4. **"Mixture" Rule" Illustrating That Testing Must Be Conducted on the Resulting Mixtures**

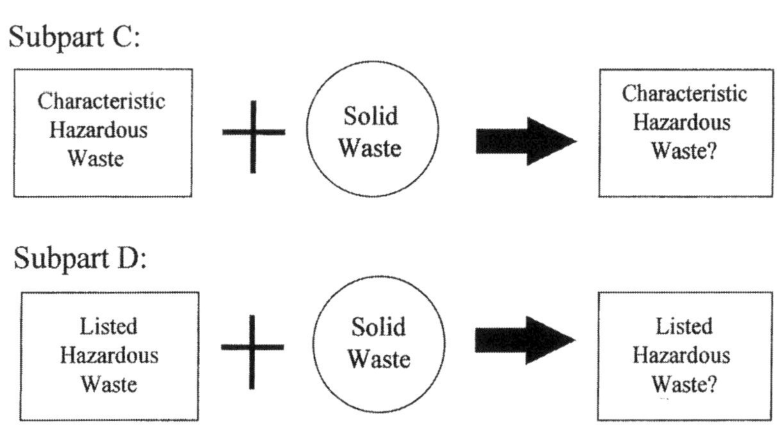

Exhibit 5. **Derived-From Rule Example**

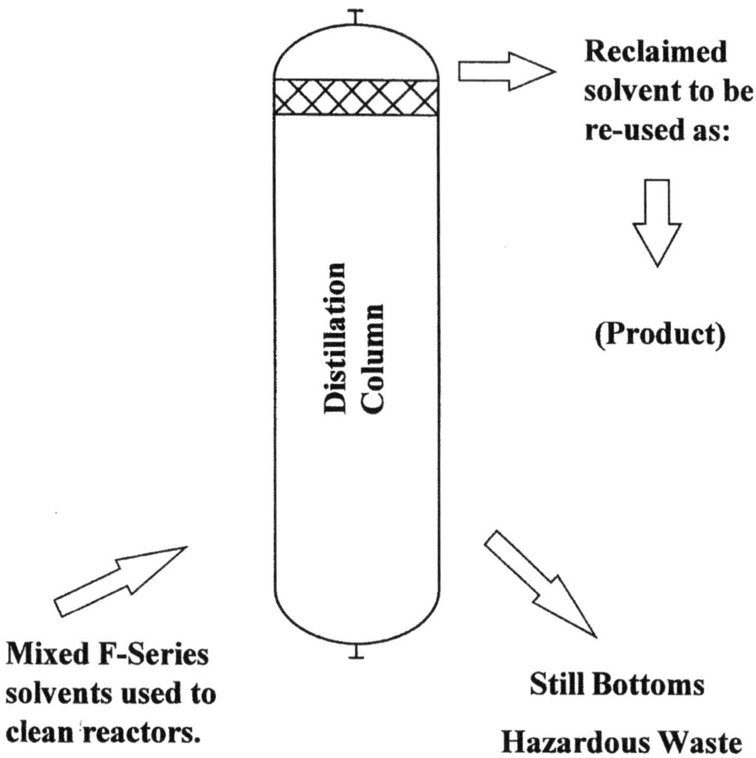

"Contained-In" Policy

Sometimes listed or characteristic wastes are spilled onto soil or contaminate equipment, buildings, or other structures. Soil (along with groundwater and sediment) is considered *environmental media*, while the equipment, buildings, and structures are considered *debris*. Examples of debris include:

- dismantled construction materials such as bricks, concrete, and wood beams;
- decommissioned industrial equipment such as pipes, pumps, and dismantled tanks;
- other discarded manufacturing objects such as personal protective equipment (PPE) (gloves, coveralls, and so on); and
- large naturally occurring objects such as tree trunks and boulders.

Environmental media and debris can be contaminated with hazardous waste in a number of ways. Environmental media can become contaminated though spills of hazardous waste or spills of product chemicals, which, when spilled, become hazardous waste. Debris can also be contaminated through spills. Most debris in the form of industrial equipment and PPE becomes contaminated with waste or product chemicals during normal industrial operations.

Waste Testing

- Audit the facility and develop a list of all wastes generated (hazardous and/or nonhazardous) and their approximate annual amounts from production areas, maintenance shops, laboratories, loading/unloading docks, wastewater treatment plants, and so on.
- Review this list with the management team to obtain agreement with the inventory. (The inventory list can also be used in pollution-prevention programs to attempt to reduce or eliminate these waste streams.)
- Meet with someone within the organization (corporate RCRA specialist) or a representative from the hazardous waste disposal company to help determine whether the wastes identified should be considered hazardous.
- Take samples and send them to a laboratory for analysis.
- From this analysis develop waste profiles to accurately describe each waste stream. Most hazardous waste disposal companies have developed a standardized waste profile sheet to gather information on waste from the generator prior to sending waste to its final destination. It is important that the information on the initial waste profile is accurate and future shipments from the same waste stream remain within the range. Sending incompatible wastes to an incinerator or fuel blender could be harmful to personnel and equipment.

Waste profile sheets are similar to material safety data sheets (MSDS) for products. Information requested on the profile will include

- general facility information;
- waste description;
- process or source that generates the waste;
- physical properties—physical state; number of phases; viscosity; color; odor; boiling, flash and melting points; pH; specific gravity; vapor pressure; total organic carbon; and BTU per pound;
- waste composition;
- constituents;
- regulatory status;
- DOT information;
- transportation requirements;
- sample status; and
- special waste-handling instructions.

Testing of waste to determine whether it's hazardous should be done by a certified laboratory that can handle RCRA waste analytical methodologies

(TCLP). EPA SW-846 outlines the testing methods for solid wastes. It is recommended that someone in your organization who is familiar with these laboratory methodologies conduct an audit of the lab facility (we will discuss auditing in Chapter 8). If this option is not available, contact other businesses or trade associations for referrals.

Land Disposal Restriction (LDR) (40 CFR 268)

Once a waste has been determined "hazardous," the next question is, "Does the waste have an LDR?" EPA has identified wastes that are prohibited from land disposal unless they meet specific treatment standards.

The LDR program consists of three main components: the *disposal* prohibition, the *dilution* prohibition, and the storage prohibition. These prohibitions restrict how wastes subject to LDR requirements are handled.

Land disposal means placement in or on the land, except in a corrective action unit, and includes but is not limited to placement in a landfill, surface impoundment, waste pile, injection well, land treatment facility, salt dome foundation, salt bed formation, underground mine or cave, or concrete vault or bunker intended for disposal purposes.

The *disposal* prohibition prohibits the land disposal of hazardous waste that has not been adequately treated to reduce the threat to human health and the environment. The criteria hazardous wastes must meet before being disposed of are known as treatment standards. These standards (found in a table in 40 CFR 268) can be either concentration levels for hazardous constituents that the waste must meet or treatment technologies that must be performed on the waste before it can be disposed.

The *dilution* prohibition prohibits dilution, such as the addition of soil or water to waste in order to reduce the concentrations of hazardous constituents. It can prohibit treatment of a waste by ineffective or inappropriate treatment methods.

The *storage* prohibition prevents the indefinite storage of untreated wastes for reasons other than the accumulation of quantities necessary for effective treatment or disposal.

Wastes that are restricted from land disposal include:

- solvent wastes specified in 40 CFR 261.31—F001, F002, F003, F004 and F005;
- dioxin-containing wastes specified in 40 CFR 261.31—F020, F021, F022, F023, F026, F027 and F028;
- the "California list" wastes as specified in 40 CFR 268.32; and
- RCRA-listed wastes as specified in 40 CFR 268.33–268.35.

The initial generator of the hazardous waste has the responsibility for determining whether the waste meets the applicable standards and communicating in writing to downstream treatment, storage, or disposal facilities regarding the LDR status of the waste. Specific LDR documentation and record-keeping requirements apply to generators and TSDFs.

The basic steps generators must take to comply with LDRs are

- to determine the significant LDR waste code(s) for the waste at the initial point of generation of the waste;
- to determine all treatment standards applicable to the waste based on the significant waste code(s);
- to determine whether the waste meets the applicable standards;
- to notify the designated TSDF in writing if the waste does not meet the applicable standards or certify that it does meet all applicable standards;
- to keep copies of all notifications, certifications, and supporting documentation used to make any of the above determinations and retain the records for at least five years.

Most hazardous waste transporters and/or brokers have developed a standardized LDR certification form that companies must review, sign, and send with a shipment. Be sure to retain a file copy.

Other Types of Waste

The Used Oil Recycling Act of 1980 was established to encourage the safe reuse of used oil and to discourage improper burning or disposal. Used oil not subject to the hazardous waste rules must be managed according to the requirements of 40 CFR 279. There are also requirements pertaining to used oil processors, transporters, marketers, and burners for energy recovery. Environmental agencies in some states (e.g., California, Massachusetts) classify used oils as hazardous wastes. Check state regulations to ensure compliance and send used oil only to reputable oil recycling companies.

Used oil (40 CFR 279), if it is recycled or reclaimed, does not have to be managed as a hazardous waste if it meets specific conditions:

- Excludes any oil that has been refined from crude oil or any other synthetic oil that has been contaminated by physical or chemical impurities.
- It is not mixed with any listed hazardous waste.
- It contains less than 1000 parts per million total halogens.
- It does not exhibit hazardous waste characteristic.

NOTE: The proper federal terminology for labels and/or markings on containers and tanks is "Used Oil."

Universal Waste Rules (40 CFR 273)

In 1995 the EPA promulgated the Universal Waste regulation to both encourage the development of recycling programs and provide significant relief from RCRA Subtitle C hazardous waste regulations for generation, collection, and handling of certain defined wastes.

Handlers of universal waste include businesses that generate and need to dispose of universal wastes (e.g., batteries, fluorescent lights); universal waste take-back programs; and universal waste collection programs.

There are two types of handlers:

- Small Quantity Handlers of Hazardous Waste (SQHUW), which accumulate less than 5000 kilograms (11,000 pounds) of universal waste at any one time.
- Large Quantity Handlers of Hazardous Waste (LQHUW), which accumulate 5000 kilograms (11,000 pounds) or more of universal waste at any one time.

The three types of wastes eligible for management as universal waste under the federal program are:

1. batteries (including lead-acid batteries);
2. certain unused pesticides; and
3. mercury-containing equipment (thermostats and lamps).

Batteries – Spent lead-acid batteries can be regulated either under the Universal Waste regulations or under 40 CFR 266.80, lead-acid battery reclamation. Batteries that are not managed under 266.80 are considered wastes upon the following:

- for a used battery, the date it is discarded
- for an unused battery, the date the handler decides to discard the battery.

The recycling process for lead-acid batteries includes cutting off the top of the battery and draining the battery of the spent acid. The positive and negative posts are pulled and melted down to be reused, while the plastic casing is shredded and also melted for reuse.

Pesticides – Only recalled pesticides under the following criteria can become universal waste:

- pesticides that are recalled under a voluntary or mandatory recall by the Federal Insecticide, Fungicide, and Rodenticide Act (FIFRA) for suspended or canceled pesticides
- suspended or canceled pesticides collected as part of a voluntary recall due to being out of compliance with FIFRA
- unused pesticides being collected as part of a state pesticide collection program.

Thermostats – Thermostats are considered solid wastes on the date a used thermostat is discarded or, for an unused thermostat, on the date the generator decides to discard it. Mercury recyclers receive old thermostats and other mercury-containing items (e.g., lamps, fluorescent bulbs, metering gauges) and feed them through a distillation process to recycle the mercury.

Universal Wastes must be managed in a way that prevents releases to the environment. Each container must be labeled, "Universal Waste—Batteries," or "Universal Waste—Pesticides" or "Universal Waste—Mercury-Containing Equipment" and the date when storage began. Accumulation of universal wastes is limited to one year from the date when storage begins. Employees must be trained annually on universal waste handling practices and emergency procedures.

The EPA's goal is to improve the recycling capabilities for these waste streams as well as to encourage generators that are not subject to the hazardous waste regulations due to their exemption status, such as households and CESQGs, to recycle their waste. Generation of universal waste does not affect a generator's hazardous waste status except in cases where waste formerly classified as hazardous is now classified as universal waste, which could allow a possible reduction in generator status.

While not federally mandated, some states now require recycling of products such as electric components; photographic wastes; aerosol cans; solvent-containing rags and wipes; treated wood; spent lamps (e.g., fluorescent or halogen bulbs); spent antifreeze; other pesticides not on the EPA list (usually you see these approved for collection on municipal waste collection days).

Summary

Chapter 3 explained how waste streams are identified and how the waste streams are then classified based on either a regulation (listed waste) or by taking a sample and analyzing it at a laboratory (characteristic). The chapter also reviewed such rules as the mixture rule and how it applies to the mixing of both hazardous and nonhazardous materials; the derived-from rule and how it applies mostly to residue remaining—for example, in the bottom of process vessels following a solvent rinse; and the contained-in rule, which focuses on equipment that has been saturated with hazardous waste—for example, laboratory equipment and decommissioned equipment such as piping and tanks. LDR was enacted to prohibit certain waste streams from land disposal. Companies now have to find alternative disposal solutions such as recycling or incineration. Finally, Chapter 3 discussed the Used Oil Recycling program and the Universal Waste rules.

Chapter 4

Management of Hazardous Waste

Chapter 4 Learning Objectives

- Understand the accumulation requirements of hazardous waste.
- Recognize the difference between storing waste streams in containers and storing them in storage tanks.
- Know how to maximize the benefits of using satellite accumulation areas.
- Understand the requirements of "less-than-90-day" storage areas.
- Know how to safely handle hazardous waste including grounding and bonding of containers.
- Determine the labels and markings required on containers during storage and prior to shipment.
- Understand the air emission control devices necessary for storage tanks.

Once it is determined that a facility is generating hazardous waste, how to manage waste in accumulation areas must be determined. RCRA provides some options for storing hazardous waste. Storage can be in containers (pails, drums, tote bins, and/or tanks,) in satellite areas, or in less-than-90-day accumulation areas. We will also discuss labeling of containers and empty container management in this chapter. Surface impoundments can also be classified as hazardous waste accumulation areas; however we will not discuss this topic in detail.

On-Site Accumulation Requirements

Regarding accumulation requirements at a facility, understanding which hazardous waste generator classification a facility falls into is very important. Remember, if a facility is classified as a CESQG—less than or equal to 100 kilograms

(220 pounds) of hazardous waste per month—there are no accumulation time requirements. If a facility is classified as an SQG—more than 100 kilograms and less than or equal to 1000 kilograms (2200 pounds) of hazardous waste per month—a facility can accumulate hazardous waste on-site for 180 days or less without a permit or for up to 270 days if it must transport, store, or treat the hazardous waste 200 or more miles away. For an LQG (greater than 1000 kilograms per month), accumulation of hazardous waste on-site must not exceed 90 days.

Let's focus on the LQG. A generator may accumulate hazardous waste on-site for 90 days or less without a permit and without having interim status provided that

- the waste is placed in containers and the generator complies with Subpart I of 40 CFR 265 (Use and Management of Containers), or the waste is placed in tanks and the generator complies with Subpart J of 40 CFR 265 (Tank Systems);
- the date upon which each period of accumulation begins is clearly marked and visible for inspection on each container;
- while being accumulated on-site, each container and tank is labeled or marked clearly with the words *Hazardous Waste*; and
- the generator complies with the requirements for owners and operators in 40 CFR 265 Subparts C and D (Preparedness and Prevention; Contingency Plan and Emergency Procedures) and Part 265.16 (Personnel Training).

An extension can be filed with the regional EPA administrator if waste cannot be removed from a facility within 90 days because of an unforeseen emergency or uncontrollable circumstances. These extensions (usually for 30 days) are judged on a case-by-case basis.

A facility may have more than one less-than-90-day accumulation area as long as it can maintain the areas and has the manpower to conduct inspections. The accumulation area(s)

- must be located in an area with limited access;
- must be clearly identified with posted signs and warning placards;
- should be away from floor drains, furnaces, or open flames; and
- should not be subject to extreme temperature changes.

The regulations focus on ensuring that

- hazardous waste containers are labeled properly;
- incompatible wastes are separated;
- the accumulation area has secondary containment;
- the accumulation area is inspected;
- written emergency procedures are up to date;
- fire control and emergency response equipment are readily available;

- site personnel involved in any aspect of handling the hazardous waste are trained; and
- arrangements have been made with local authorities regarding emergency response activities.

We'll discuss emergency response activities in Chapter 6.

Hazardous Waste Storage Tanks (40 CFR 265, Subpart J)

Bulk storage tanks can also be used as less-than-90-day hazardous waste accumulation tanks. The 90-day time frame starts as soon as the first drop of hazardous waste enters an empty tank. *Empty* is defined as "when the contents have been drained to the fullest extent possible." As a best practice, it is generally recommended that generators leave the tank drainage system open until a steady continuous flow has ceased.

There are four parts to hazardous waste storage tank regulations:

1. **Integrity Assessments (40 CFR 265.193 (a)):** An integrity assessment must be performed, documented, and certified by an independent professional engineer (PE). The certification must prove the tank system has sufficient structural integrity to store hazardous waste.
2. **Detection and Containment of Releases (40 CFR 265.193 (b)–(p)):** Appropriate secondary containment for tanks and ancillary equipment (e.g., pumps, flanges, valves) must be designed, installed, and operated to prevent migration of any wastes and liquids and must be capable of detecting and collecting releases of liquids until the material is recovered. A leak detection system must be in place that is capable of detecting tank failures within 24 hours.
3. **Operation and Inspection (40 CFR 265.194 and .195):** The tank system must use appropriate controls and practices to prevent spills and/or overflows. At minimum the system must include spill prevention controls (e.g., check valves, dry-disconnect couplings) and overfill protection controls (e.g., level sensors, high-level alarms, automatic feed cutoff or bypass).

   Aboveground portions of the tank and system must have a documented visual inspection performed daily. Piping, welded flanges, joints and connections, and pumps must be visually inspected unless the ancillary equipment has full secondary containment (e.g., double-walled piping, sealless pumps and valves). Also, a daily operating record of what is stored in the tank must be maintained.
4. **Closure and Post-Closure (40 CFR 262.34 (a)(1) and 265.111):** If a tank is no longer used for less-than-90-day hazardous waste storage, all waste residues must be removed through an appropriate decontamination process. The tanks and ancillary equipment must be removed according

to the Closure Performance Standard in 40 CFR 265.111 and sent out as hazardous waste.

Satellite Accumulation Areas – 40 CFR 262.34 (c)(1)

One feature of the RCRA regulation that can work to a generator's benefit is satellite accumulation—the accumulation of up to 55 gallons of hazardous waste or one quart of acutely hazardous waste (P listed) in containers at or near the point of generation where wastes initially accumulate, which is under the control of the operator of the process generating the waste, providing that

- the generator complies with 40 CFR 265.171, 265.172, and 265.173 regarding management of containers (e.g., containers in good condition, containers compatible with materials, containers always closed during storage, proper container handling); and
- the containers are marked with the words *Hazardous Waste* or with other words that identify the contents of the container (e.g., *Waste Styrene and Water*).

Once the 55-gallon container of hazardous waste or one quart of acutely hazardous waste has been filled, the generator must move the container to the facility's less-than-90-day accumulation area (180-day for SQG) within three calendar days and place the accumulation start date on the container label. The benefit of having a satellite accumulation area is that a facility can accumulate hazardous wastes in the satellite area for more than 90 days.

An example of a satellite accumulation area is an area where a quality assurance lab accumulates one quart of mixed lab waste per eight-hour shift. At the end of the shift, a technician transfers the contents of the hazardous waste storage can to a 55-gallon drum stored outside the lab area.

Another example is a production facility with a raw material off-loading area; during disconnect of the rail cars or tank trucks, a few drops of material fall into a spill pan, and the operator pours the contents into a grounded 55-gallon accumulation drum.

In both of these scenarios, if the drums were transferred to the less-than-90-day accumulation area following transferring from the lab can and unloading the spill pan to the drum, shipment of both of these drums would occur within 90 days. With the satellite accumulation allowance, both drums are permitted to be filled without penalty during normal operations that could take more than 90 days.

A few safety precautions must be followed when transferring hazardous wastes from a small container to a 55-gallon drum:

- Ensure that personnel have been trained on the use of applicable personal protective equipment.
- Ensure that proper ventilation is provided.

***Exhibit 6a* and *6b*. Photos of a Satellite Hazardous Waste Accumulation Facility in Operation**

(a)

(b)

- Ensure that the transfer of flammable materials is conducted using proper grounding and bonding procedures (satellite accumulation containers should be grounded).
- Ensure that spill containment equipment is available nearby.

One word of caution with regard to satellite areas located at or near the point of generation and under an operator's control: Most regulatory inspectors won't give a definitive interpretation of "at or near the point of generation," because this is a subjective call. A company's definition and the inspector's definition may differ, so as a rule of thumb, keep the distance from the work area to the satellite area within reasonable walking distance (e.g., outside the door of a laboratory). Not only does this keep the responsibility on the person filling the drum, it also reduces the liability associated with spills.

Safe Handling of Hazardous Waste

Frequently, the first way hazardous waste is handled is moving it. Whether a worker is collecting small containers of waste for transfer to larger containers, moving a large container around the worksite and doing transfers at another location, or just moving containers to on-site storage, there are several things to consider.

The first consideration is personal protective equipment (PPE). What the worker must wear depends on the potential danger from exposure to what is being moved. A company's safety and industrial hygiene personnel should be consulted to ensure that proper PPE is selected. Personnel also need to be properly trained to wear the appropriate PPE.

Splashes are probably the most common form of hazardous waste exposure. When opening a drum, always open the small bung first, and unscrew it slowly. If there is any pressure in the drum, it should be able to vent while the cap is still partially threaded into the lid. Always use a funnel when pouring from one container to another, and ensure that the containers are grounded and/or bonded.

Two common type of containers used in hazardous waste storage are 55-gallon drums and 5-gallon cans. Some of these cans are self-closing to prevent spills if they tip over and self-venting to prevent a dangerous buildup of explosive vapors. Self-venting cans are required by law for certain flammable substances. The facility shown in Exhibit 7 was designed to protect workers, contain spills, and protect the 55-gallon drum. Exhibit 8 shows a drum standing on a containment vessel.

To move small waste containers, use a cart that has the capacity for secondary containment. If a container spills, the liquid would spill into the cart and not on the ground or on a person.

Prior to moving a 55-gallon drum, check to make sure that both bung caps are tight. Always check the drum for rust, dents, and/or leaks, especially if it has been sitting out in the elements for a period of time. If there is any evidence of damage to the drum, do not move it; consult with the appropriate personnel within the company to determine what needs to be done. To move a 55-gallon drum any distance, use either a drum dolly with some type of restraint or a forklift.

Falling drums are one of the most common causes of injuries. Not only does the drum have a chance of landing on a worker, it can also pin someone, and if it ruptures it can expose people to the contents of the drum. If a drum starts to fall, let it go.

Exhibit 7. **Example of a Hazardous Waste Satellite Accumulation Container**
The container is grounded, spill containment is in place, and a bumper pole is in place to protect the pallet. A pouring funnel is not shown. (Photo courtesy of BASF Corporation, South Brunswick, NJ.)

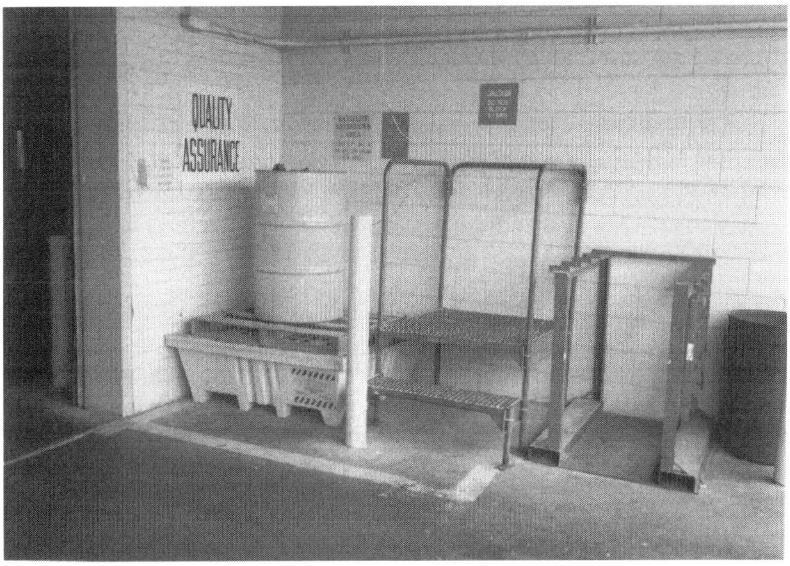

Exhibit 8. **Example of a Used Oil Satellite Accumulation Container**
Spill containment is in place and the pouring funnel is locked for control of the container. (Photo courtesy of BASF Corporation, South Brunswick, NJ)

Vapor or particulate releases are other concerns when handling hazardous waste. These releases can harm people in more ways than one. Inhalation of hazardous waste can poison people or irritate the eyes, nose, mouth, or lungs. Many vapors are flammable, so there is the potential of fire or explosion. Some vapors have the ability to corrode pipes, sprinkler heads, or equipment. Work areas where hazardous wastes are handled, and specifically areas where they are transferred, should have adequate ventilation and/or an exhaust system that draws airborne contaminants away from workers and out of the area. For certain types of waste, respiratory protection may be required.

Another hazard is the mixture of incompatible wastes; it is important to keep them separated. This is called *segregation and separation.* Remember reactivity? Some chemicals—such as acids and wastes containing sulfide—can become explosive when mixed together. Never mix two hazardous wastes together unless you are certain they are compatible. Also, be sure the container is compatible with the contents.

Finally, be cognizant of the importance of grounding and bonding. Ensure that all stored metal containers are properly grounded. The *storage* container must be grounded, while the *receiving* container (usually a 1- to 5- gallon container) must be bonded to the storage container using bonding straps or cables to eliminate the potential generation of a static discharge (Exhibit 9). This concept also applies to transfer of liquid waste materials between plastic containers to reduce the potential of static electricity.

Safe Storage

While safe storage is not much different from safe handling, here are a few recommended practices for the safe storage of hazardous waste:

- Keep the area well ventilated.
- Be sure the drums are grounded (if applicable).
- Keep drums away from heat and flames.
- Protect drums from extreme cold.
- Store flammable and corrosive materials in approved containers.
- Keep drums away from floor drains.
- Follow rules for proper segregation and separation of hazardous materials.
- Put nothing into storage until it has been properly and completely marked and labeled.
- Make sure markings and labels are visible.

Containers (40 CFR Subpart I 264.170 – .179)

This section describes the management of containers for storing hazardous wastes. The boxes, cartons, drums (fiber, plastic, or metal), overpacks, salvage

***Exhibit 9.* The Use of Both Bonding and Grounding Equipment (Illustration courtesy of the Stewart R. Browne Company.)**

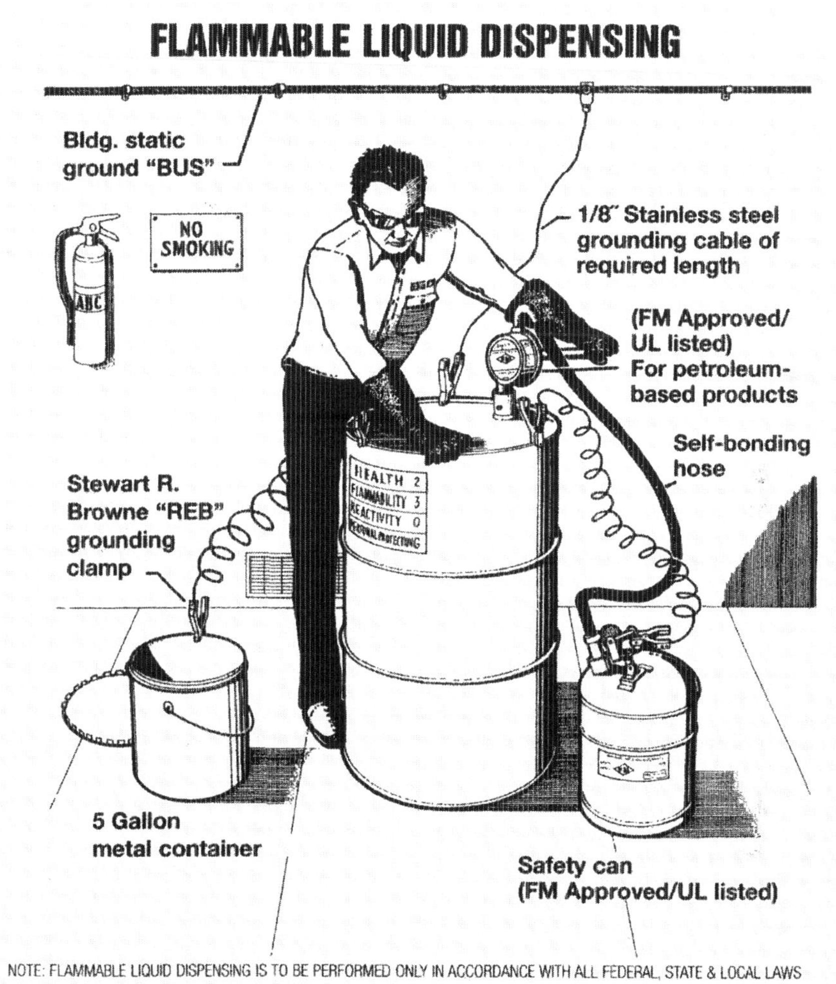

drums, and tote bins that hazardous waste is stored in for shipment to a waste disposal facility must meet the Department of Transportation (DOT) HM-181 packaging specifications for shipping and labeling hazardous waste. Check with your company's supplier to ensure that the containers meet DOT specifications.

The regulations state that the hazardous waste container must be in good condition (no severe rusting or apparent structural damage), or if not, the generator is responsible for transferring the material prior to shipping the container off-site.

The generator must ensure that the container or container liner is compatible with the material to be placed in the container to prevent any reactions. The generator must ensure that the container remains closed (e.g., bungs and lids tightened) during storage, except when it is necessary to add or remove waste.

On a weekly basis, the generator must inspect the area where the containers are stored, looking for leaking containers and deterioration of containers and the containment system caused by corrosion and other factors. Always document inspections for, at a minimum, internal audits and agency visits.

Containers must be stored in an area where the containment system has been designed and operated as follows:

- A base must underlie the containers which is free of cracks or gaps and sufficiently impervious to contain leaks, spills, and accumulated precipitation until leaked material is detected and removed.
- The base must be sloped or the containment system designed and operated to drain and remove liquids resulting from leaks, spills, or precipitation unless the containers are elevated or are protected from contact with accumulated liquids.
- The containment system must have sufficient capacity to contain 10 percent of the volume of all containers or the volume of the largest container, whichever is greater;
- Run-on (rain water, snow melt) into the containment system must be prevented unless the collection system has sufficient excess capacity to contain any run-on that may enter the system.
- Spilled or leaked waste and accumulated precipitation must be removed from the sump or collection area in as timely a manner as is possible to prevent overflow of the collection system.

Containers holding ignitable and reactive wastes must be located at least 50 feet from the facility's property line, and precautions must be taken for storage of incompatible waste streams.

Empty Container Management (40 CFR 261.7)

A container is considered empty if

- all wastes have been removed using common practices, and
- the total waste remaining is less than 3 percent by weight if the container is 110 gallons or less, or 0.3 percent if the container is more than 110 gallons, or
- no more than 1 inch of residue remains.

A container (or inner liner) that held an acutely (P-listed) hazardous waste must be triple rinsed. Triple rinsing is not considered treatment; however, the rinsate is a hazardous waste. Residues from the empty containers, after rinsing, are not considered hazardous waste.

An empty container is "closed" if all closures on the container (bungs, lids, rings, bolts) are in place and no holes or punctures are present. Labels and markings should be in place as if the drum contained its original contents. These labels include product labeling, hazard communication labeling, and DOT labeling.

Intermediate bulk containers (IBCs), otherwise known as tote bins, are becoming popular as hazardous waste transfer containers. Most IBCs can hold between 200 and 400 gallons of material, which makes them easier to handle than four 55-gallon drums. However, before beginning to store hazardous waste in IBCs, make sure the hazardous waste disposal company can handle tote bins and is willing to transfer or incinerate them. Make sure the tote bins are compatible with the waste that will be stored in them and that they are DOT HM-181 approved. (See Table 1 for a comparison of storage requirements.)

Labeling and Marking (40 CFR 262.31 and 262.32)

Prior to transporting and offering hazardous waste for transportation off-site, a generator must label each package (e.g., box, drum) in accordance with the applicable Department of Transportation regulations on packaging materials under 49 CFR 172. These labels reflect the DOT hazard classes (e.g., corrosive, flammable liquid, oxidizer, organic peroxide).

Also, prior to shipping, the generator must mark each container of 110 gallons or less with the following words and information displayed in accordance with the requirements of 49 CFR 172.304:

> HAZARDOUS WASTE Federal Law Prohibits Improper Disposal. If found contact the nearest police or public safety authority or the U.S. Environmental Protection Agency.
>
> Generator's name and address:
>
> Manifest document number:

Standardized hazardous waste labels (commonly yellow in color, but that is not required) can be purchased and even customized with your company's name, address, and telephone number (Exhibit 10).

RCRA Air Emission Requirements (40 CFR 264 and 265, Subparts AA, BB, and CC)

In December 1996, the EPA published a final ruling amending the RCRA rules on volatile organic air emissions for tanks, surface impoundments, and containers.

Subpart AA (40 CFR Part 265.1030)

Subpart AA requires process vents associated with distillation, fractionation, thin-film evaporation, solvent extraction, or air or steam stripping operations that manage hazardous waste with a volatile organic (VO) concentration of at least 10 parts per million by weight (ppmw) to be connected to a control

Table 1. Comparison of SQG and LQG Storage and Inspection Requirements

Container Management Standards for Generators

Section of Part 265 Subpart I		Satellite Accumulation Area	180-day area (SQG)	90-day area (LQG)
265.171	Condition of containers	**YES**	**YES**	**YES**
265.172	Compatibility of waste with containers	**YES**	**YES**	**YES**
265.173 (a)	Keep closed, except when add/remove waste	**YES**	**YES**	**YES**
265.173 (b)	Handle containers to avoid ruptures and leaks	**no**	**YES**	**YES**
265.174	Inspections	**no**	**YES**	**YES**
265.176	Special requirements of ignitable or reactive wastes	**no**	**no**	**YES**
265.177	Special requirements for incompatible wastes	**no**	**YES**	**YES**
265.178	Air emission standards	**no**	**no**	**YES**

Exhibit 10. **Sample Hazardous Waste Label**

HAZARDOUS WASTE

FEDERAL LAW PROHIBITS IMPROPER DISPOSAL.

IF FOUND, CONTACT THE NEAREST POLICE, OR PUBLIC SAFETY AUTHORITY, OR THE U.S. ENVIRONMENTAL PROTECTION AGENCY.

GENERATOR INFORMATION:

NAME ______

ADDRESS ______ PHONE ______

CITY ______ STATE ______ ZIP ______

EPA ID NO. / MANIFEST DOCUMENT NO. ______ / ______

ACCUMULATION START DATE ______ EPA WASTE NO. ______

D.O.T. PROPER SHIPPING NAME AND UN OR NA NO. WITH PREFIX

HANDLE WITH CARE!

STYLE WM6

Printed by Labelmaster, An American Labelmark Co., Chicago, IL 60646 (800) 621-5808

device (i.e., thermal oxidizer) that effectively reduces the organic vapors routed to it.

Subpart BB (40 CFR Part 265.1050)

Subpart BB applies to certain pieces of equipment that contain or contact hazardous waste with a VO concentration of at least 10 percent by weight. Regulated equipment examples are pumps, compressors, open-ended lines, valves, flanges, and pressure-relief devices. This equipment must be marked or tagged for identification. The equipment is then subject to a leak detection and repair (LDAR) program that includes visual inspections and/or EPA Test Method 21 monitoring. Subpart BB requires leaks, as defined by the regulation, to be repaired within a specified time period.

Subpart CC (40 CFR Part 265.1080)

Subpart CC requires units that manage hazardous wastes with an average VO concentration greater than or equal to 500 ppmw at the point of waste origination that aren't specifically exempt from the requirements to be designed and operated to meet air emission control requirements. This can be accomplished by containing vapors within the unit and in some cases routing vapors to a control device. Subpart CC requires the control device to remove or destroy at least 95 percent of the vapors vented to it. There are specific technical requirements for the control device depending on the type of unit selected.

Level of Air Emission Controls

Subpart CC establishes levels of air emission controls for tanks and containers. These levels are specific to the size and design of the unit as well as the vapor pressure and composition of the waste to be managed in the unit.

Tank Controls (40 CFR Part 265.1085)

Tanks that receive hazardous waste with a VO concentration of 500 ppmw (parts per million by weight) at the point of origination that have not been exempted are subject to Subpart CC air emission controls. Level 1 controls may be used if

1. the tank does not exceed the maximum organic vapor pressure (MOVP) specified in Table 2;
2. the waste in the tank is not heated to a temperature greater than the temperature at which the MOVP is determined (this is to ensure that organic vapors are not freely vented to the atmosphere); and
3. no waste stabilization occurs in the tank.

If these criteria are met, the hazardous waste may be placed in a Level 1 fixed-roof tank. Level 2 tanks are fixed-roof tanks with closed vent systems and controls, floating-roof tanks, pressure tanks, and tanks inside an enclosure. The MOVP may be determined by either testing or knowledge. Table 3 summarizes the three control levels.

A fixed-roof tank is a tank whose cover is part of the structural design and forms a continuous barrier over the entire surface of the tank. Closure devices on tanks include caps, hatches, lids, plugs, seals, valves, and other types of fittings that block an opening in a fixed-roof cover so that when they are closed, air emissions to the atmosphere are prevented or reduced.

Tank covers must be closed at all times except when performing routine inspections, maintenance, or other normal operations; and when removing accumulated sludges or other residues from the bottom of the tank. When operations are completed, the covers should be promptly closed.

Table 2. Maximum Vapor Pressure Limits for Tanks Used to Store Hazardous Wastes

Tank Size	Vapor Pressure Limit Must Be
>40,000 gal	<0.75 psi
>20,000 and <40,000 gal	<4.0 psi
<20,000 gal	<11.1 psi

Fixed-roof tanks that meet Level 1 controls may be equipped with *conservation vents*. Conservation vents are pressure/vacuum-relief devices designed to vent to the atmosphere during normal operations to maintain internal tank pressure.

Subpart CC allows the use of safety relief devices on tanks. "Safety relief devices" refers to a closure device such as a pressure relief valve, frangible disc, fusible plug, or any other device used to prevent physical damage to or permanent deformation of a unit or its air control equipment during unsafe conditions resulting from an accidental or emergency event. Such devices must remain closed during normal operations.

Container Controls (40 CFR Part 265.1087)

Containers that have a design capacity of less than 0.1 cubic meters (approximately 26 gallons) are exempt from Subpart CC. Other containers are subject to one of three levels of air emission controls (Table 4). Level 1 containers have a design capacity of greater than 26 gallons and less than 119 gallons or greater than 119 gallons but not in light material service. "Light material service" means that the vapor pressure of the material managed in the container is greater than 0.3 kilopascals (.043 pounds per square inch) at 20°C and the total concentration of pure organics at this vapor pressure is equal to or greater than 20 percent by weight. Level 1 containers must meet one of the following control options:

- the applicable DOT standards (49 CFR 173, 178, 179 and 180)
- the covered container standards of the original subpart CC rule (no visible holes or gaps).

Containers that have a design capacity of greater than 119 gallons that are in light material service are Level 2 containers and must meet one of the following control options:

- the applicable DOT standards (49 CFR 173, 178, 179 and 180)
- the "no detectable organic emission" standard when tested by EPA Test Method 27 of 40 CFR 60, Appendix A
- the "vapor-tight standard" when tested by EPA Test Method 27 of 40 CFR 60, Appendix A.

Table 3. Summary of Tank Air Emission Control Requirements (40 CFR Part 265.1085)

Control Level	Control Requirements	Inspection & Monitoring Requirements	Recordkeeping Requirements
LEVEL 1	Fixed roof with no capture or destruction of vapors Tight-fitting covers & closures—safety devices & conservation vents allowed	Initial and annual visual inspection of cover, closures, defects Maintain written plan & schedule for inspections of each tank	Records of tank ID, MOVP, inspection dates, defect repair
LEVEL 2	Fixed roof with internal floating roof with proper seals Hatches & openings kept closed—operated to design requirements	Visual inspect seals, gaskets, membranes, covers initially, after each emptying Written plan & schedule —notify EPA prior to inspection	Records of tank ID, dismensions, inspectino dates, defect repair
	External floating roof with proper seals Hatches & openings kept closed—operated to design requirements	Visually inspect covers & measure gaps between tank wall & seals initially & annuall (5 years for primary seals) Written plan & EPA notification	Records of dimensions, inspection dates, defect repair

Table 3. Summary of Tank Air Emission Control Requirements (40 CFR Part 265.1085) (contined)

Control Level	Control Requirements	Inspection & Monitoring Requirements	Recordkeeping Requirements
LEVEL 2	Fixed roof venting to a closed vent system and control device Closed vent system control—95% reduction of organics from control device	Initial & annual visual inspection of covers, closures, defects—EPA Method 21 monitoring on closed vent sytem Written plan & schedule to inspect	Records of tank ID, inspection dates, defect repair, semiannual PM on control device, malfunctions, ID CEMS or CMS
	Pressure tank that operates with no detectable emissions during normal operations	Initial & annual visual inspection of covers, closures, defects Written plan & schedule to inspect	Records of tank ID, inspection dates, defect repair, MOVP
	Enclosure vented to an enclosed combustion device Closed vent system and control device achieving 95% reduction	Initial & annual visual inspection of covers, closures, defects—EPA Method 21 monitoring on closed vert system Written plan & schedule to inspect	Records of tank ID Inspection dates, defect repair, semiannual PM on control device, malfuctions, MOVP, enclosure calculations

Table 4. Summary of Container Emission Control Requirements (40 CFR Part 265.1087)

	Container Size	Regulated Material	Container Storage Requirements	Loading Requirements	Inspection & Monitoring Requirements	Recordkeeping Requirements
LEVEL 1	≤ 0.1 m^3 (approx. 26 gal)	All	None	None	None	None
	> 0.1 m^3 and ≤ 0.46 m^3 (approx. 26.4 and 119 gal)	≥ 500 ppmw	Alternatives: 1. Meets NOT Standards. 2. Cover w/no visible gaps or holes 3. Vapor suppressing barrier Maintain each closure/cover in closed position except to access waste	Container can remain open during continuous transfers Container to be closed between batch transfers ≥ 15 min.	Initial visual inspection of container cover/closures Annual visual Inspection if container remains on-site for ≥ 12 monts. Repair container defects within 5 days	Light material service records, if applicable Visual inspection records not required
	> 0.46 m^3	≥ 500 ppwm & NOT in light material service				

The use of approved DOT containers will satisfy either Level 1 or Level 2 controls.

Level 3 containers are those that have a design capacity greater than 26 gallons that are used for treatment of a hazardous waste by a stabilization process. Organic vapors from Level 3 containers must be routed to a control device when, during stabilization, waste in the container is exposed to the atmosphere.

Conservation vents and safety devices are allowed on containers. Safety devices should remain in a closed position during normal operation. Covers and closures do not need to be in place and secure on RCRA empty containers.

Subpart CC control options for surface impoundments include the use of either a cover that vents vapors to a control device or a floating membrane cover.

Container Loading Requirements

There are requirements under Subpart CC for transferring hazardous waste in or out of Level 2 containers. Specifically, the rule requires loading and unloading of wastes to be done in a manner that minimizes the exposure of the waste to the atmosphere, taking into consideration the composition of the waste and good engineering and safety practices. The rule cites the following practices as examples that would meet these requirements:

- submerged-fill pipe or bottom loading
- vapor-balancing or vapor-recovery systems
- top loading with transfer line purging system.

It is important to note that the EPA has clarified its intent to allow the practice of vacuum loading whereby venting vapors to the atmosphere during waste transfer operations is permissible.

Surface Impoundment Controls (Subpart CC)

Subpart CC, Air Emission Standards for Tanks, Surface Impoundments and Containers, specifies air emission control options for surface impoundments includes the use of either a cover that vents vapors to a control device or a floating membrane cover (264.1087).

Summary

Chapter 4 discussed how to accumulate hazardous waste and how long it may be accumulated. The chapter also outlined what is required for each type of accumulation area (satellite, less-than-90-day, 180-day). Containers (drums) must be labeled and properly secured, while storage tanks must meet integrity testing requirements based on engineering protocols. Satellite accumulation can benefit a generator if used correctly and effectively. Safe handling of containers, including transfer of materials and grounding and bonding of pails, drums, and storage tanks, was discussed in relation to the personal safety of operators. Empty container management described when a container or storage tank is considered empty. Labeling and marking of containers being stored at the generator location or transported over the road under DOT guidelines was also discussed. Finally, the required technology to control volatile organic air emissions of materials stored in storage tanks and 55-gallon drums was demonstrated through a series of charts and tables.

Chapter 5

Hazardous Waste Manifests

Chapter 5 Learning Objectives

- Be able to complete and distribute the hazardous waste manifest (shipping document) that is required to accompany all transported loads.
- Determine who must retain and receive a copy of the manifest.
- Understand how to handle a rejected or missing load.

The manifest is the tracking document behind the entire RCRA cradle-to-grave concept. It is the official DOT shipping document that tracks a hazardous waste shipment from the time it is loaded on the hazardous waste transporter's vehicle until it is disposed of (incinerated, fuel blended, landfilled, or recycled, for example). The manifest carries with it legal liability for the person from the facility who signs it.

Distribution of the manifest copies is important as well to ensure that all applicable parties (shown in Exhibit 11) have received their respective copy in a timely fashion. Since the federal regulations changed on September 5, 2006, the Uniform Hazardous Waste Manifest is now used for all shipments throughout the United States. Individual states no longer have their own manifests. The manifest consists of at least six copies so that the generator, each transporter, and the owner or operator of the designated TSDF facility each has a copy for its records. When the hazardous waste shipment has been accepted at the TSDF, one signed manifest copy is then returned to the generator, while another is sent to the state environmental agency (if applicable—check with the state agency to find out if it wants a copy of the completed manifest). A key to completing the manifest is provided in Exhibit 12, and a completed manifest can be found in the appendix.

Exhibit 11. **Distribution of a Hazardous Materials Manifest**

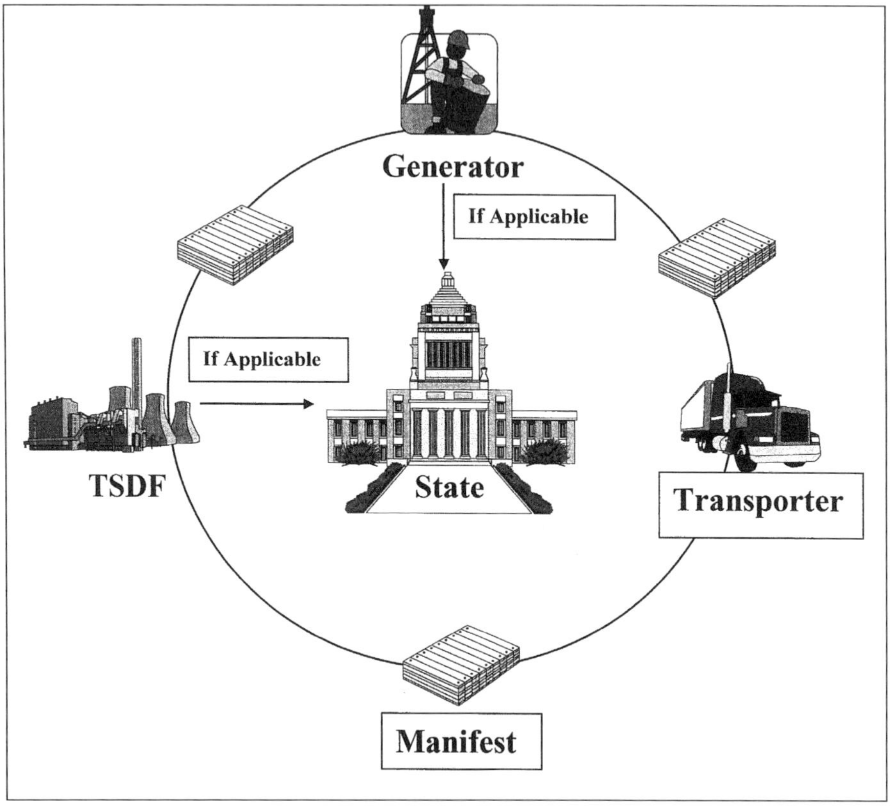

Exhibit 12. **Instructions for Completing a Hazardous Waste Manifest**

Identifier	Explanation
A	Generator EPA ID #
B	Generator's name, address and telephone
C	Name of transporter or transporters & EPA ID #
D	Name, address, EPA ID # of TSDF facility receiving the waste
E	U.S. DOT Descriptions must indicate the following information in the order listed:

1. Proper shipping name
2. Hazard class
3. I.D. number (UN or NA number)
4. Packaging group:
   a. I - great danger
   b. II - medium danger
   c. III - minor danger

5. EPA waste code
6. NAERG number (for compliance with the Emergency Response Communication Standard)

F If the container material has a Reportable Quantity applicable the letters "RQ" must be included in the Shipping Description (either in the HM (9a) column or before/after the basic Shipping Description).

If the word "waste" is not part of the basic description, it must appear before the description as listed in 49 CFR 172.101.

The two primary constituents causing a material to be hazardous for Shipping Names utilizing a N.O.S. (Not Otherwise Specified) description, must be entered in parenthesis in the DOT Shipping Description (after the Packaging Group number, or after the Shipping Name).

G A current manifest must be used.

H List the number of manifest pages. There may be multiple pages if a Continuation Sheet is used. As an alternative, multiple manifests for a shipment may be prepared if all items do not fit on a single form. Each manifest must have a unique number.

I 24 hour Emergency Response Phone #

J Section 10: Type of Containers

DM metal drums, barrels, kegs
DW wooden drums, barrels, kegs
DF fiberboard or plastic drums, barrels, kegs
TP tanks, portable
TT cargo tanks (tank trucks)
TC tank cars (rail cars)
DT dump trucks
CY cylinders
CM metal boxes, cartons, cases (includes roll-offs)
CW wooden boxes, cartons, cases
CF fiber or plastic boxes, cartons, cases
BA burlap, cloth, paper or plastic bags
HG hopper or gondola cars

K Section 11: Quantity of Containers

L Section 12: Units of Measure

G gallons (liquids only)
P pounds
T tons (2,000 lbs.)
Y cubic yards
L liters (liquids only)

K	kilograms
M	metric tons (1,000 kg)
N	cubic meters

M Federal and/or State waste codes

N Must be signed/dated by generator

O Must be signed/dated by the transporter

P Must be signed/dated by receiving TSDF facility.

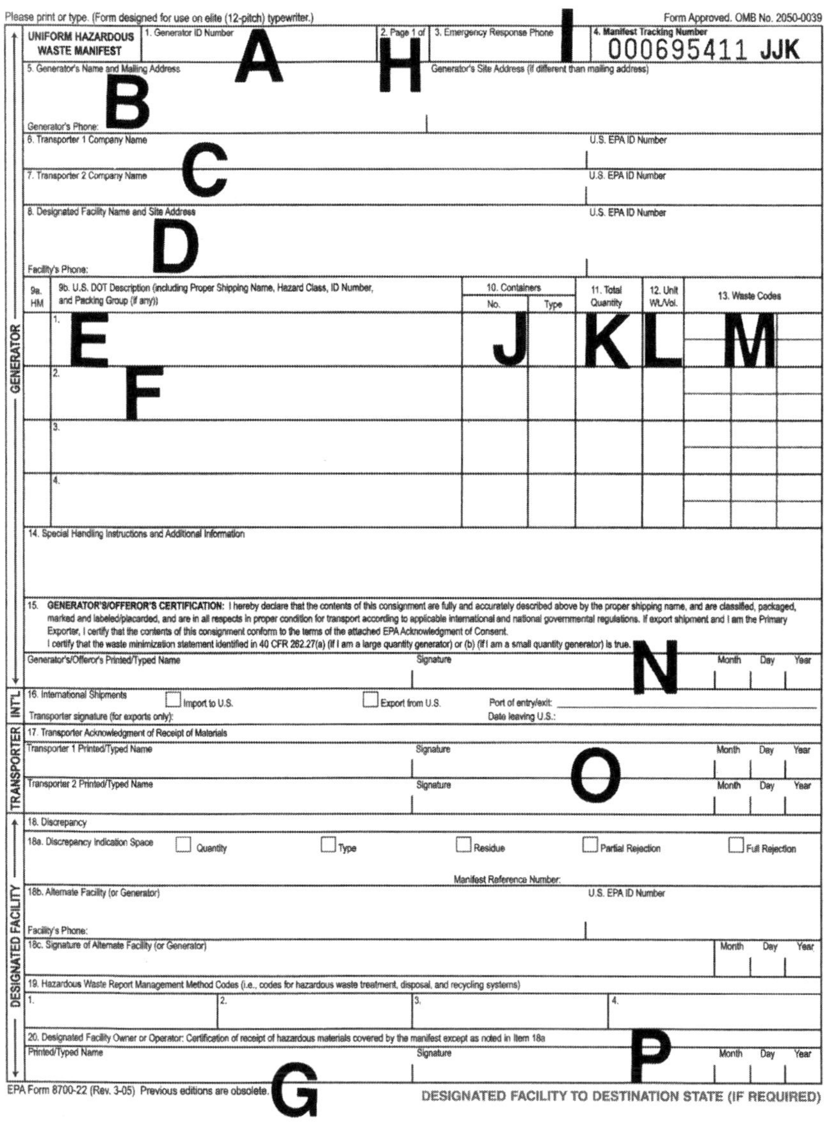

Please print or type. (Form designed for use on elite (12-pitch) typewriter.) Form Approved. OMB No. 2050-0039

UNIFORM HAZARDOUS WASTE MANIFEST | 1. Generator ID Number | 2. Page 1 of | 3. Emergency Response Phone | 4. Manifest Tracking Number 000695411 JJK

5. Generator's Name and Mailing Address | Generator's Site Address (if different than mailing address)

Generator's Phone:

6. Transporter 1 Company Name | U.S. EPA ID Number

7. Transporter 2 Company Name | U.S. EPA ID Number

8. Designated Facility Name and Site Address | U.S. EPA ID Number

Facility's Phone:

9a. HM | 9b. U.S. DOT Description (including Proper Shipping Name, Hazard Class, ID Number, and Packing Group (if any)) | 10. Containers No. Type | 11. Total Quantity | 12. Unit Wt./Vol. | 13. Waste Codes

1.
2.
3.
4.

14. Special Handling Instructions and Additional Information

15. GENERATOR'S/OFFEROR'S CERTIFICATION: I hereby declare that the contents of this consignment are fully and accurately described above by the proper shipping name, and are classified, packaged, marked and labeled/placarded, and are in all respects in proper condition for transport according to applicable international and national governmental regulations. If export shipment and I am the Primary Exporter, I certify that the contents of this consignment conform to the terms of the attached EPA Acknowledgment of Consent.
I certify that the waste minimization statement identified in 40 CFR 262.27(a) (if I am a large quantity generator) or (b) (if I am a small quantity generator) is true.

Generator's/Offeror's Printed/Typed Name | Signature | Month Day Year

GENERATOR

INT'L 16. International Shipments ☐ Import to U.S. ☐ Export from U.S. Port of entry/exit:
Transporter signature (for exports only): Date leaving U.S.:

TRANSPORTER 17. Transporter Acknowledgment of Receipt of Materials
Transporter 1 Printed/Typed Name | Signature | Month Day Year
Transporter 2 Printed/Typed Name | Signature | Month Day Year

DESIGNATED FACILITY 18. Discrepancy
18a. Discrepancy Indication Space ☐ Quantity ☐ Type ☐ Residue ☐ Partial Rejection ☐ Full Rejection
Manifest Reference Number:
18b. Alternate Facility (or Generator) | U.S. EPA ID Number
Facility's Phone:
18c. Signature of Alternate Facility (or Generator) | Month Day Year

19. Hazardous Waste Report Management Method Codes (i.e., codes for hazardous waste treatment, disposal, and recycling systems)
1. | 2. | 3. | 4.

20. Designated Facility Owner or Operator: Certification of receipt of hazardous materials covered by the manifest except as noted in Item 18a
Printed/Typed Name | Signature | Month Day Year

EPA Form 8700-22 (Rev. 3-05) Previous editions are obsolete.

DESIGNATED FACILITY TO DESTINATION STATE (IF REQUIRED)

The generator must designate on the manifest one facility that is permitted to handle the waste described. The generator can also designate one alternate facility in case an emergency prevents delivery of the waste. If the transporter cannot deliver the waste to the primary or alternate facility, the generator should designate another facility or instruct the transporter to return the waste to the originating facility.

The generator must sign the manifest certification by hand, obtain the handwritten signature of the initial transporter and date of acceptance on the manifest, and retain three copies. The generator keeps one copy, one copy is forwarded to the generator's state agency, and one copy is forwarded to the hazardous waste destination's state agency (only if applicable). The generator must then give the transporter the remaining copies of the signed manifest.

The dates next to the signatures are critical. There are limitations governing how long a waste can be in transit. If a generator does not receive a signed, completed manifest from the TSDF within 35 days of shipment, the generator must track the waste. If the waste is not found within 45 days for a large- quantity generator or 60 days for a small-quantity generator, an exception report must be provided to either the EPA or the state agency. Some states have more stringent requirements than others. Establish a tickle file to flag waste shipment manifests not returned within a certain time frame (for example, 20 days). This will allow time to check with either the transporter or the TSDF to determine the reason for the delay.

Sometimes TSDFs reject shipments because the manifest doesn't have the right paperwork attached. For example, virtually all liquid hazardous wastes are banned from landfills until they are rendered nonhazardous. When disposing of one of these materials, land ban certificates must be attached to the manifest.

Another reason for a TSDF to not accept a waste shipment is that it discovers after analysis of a sample that the waste does not meet the specifications of the waste profile agreed to prior to shipment. This can cause the TSDF treatment and/or catastrophic (fire, explosion) problems.

The key to completing the manifest is to understand the DOT regulations (49 CFR 172) for proper shipping names. Most hazardous waste transportation company representatives can be called on for help in filling out a manifest prior to shipment. This information is gathered from the waste profile already on record with the waste disposal company.

Another key resource used to complete the manifest is the DOT's North American Emergency Response Guidebook (NAERG). Each hazardous material shipped on the highways and railroads in North America is affixed with a United Nations (UN) or North America (NA) number. These are the numbers you see on the DOT class placards on the sides of rail cars and trucks (e.g., methyl ethyl ketone (MEK) is UN1193, sulfuric acid is UN1830).

Exhibit 13 provides a quick guide to UN labels for both solids and liquids.

Exhibit 13. **Explanations of U.N. Labels for Liquids and Solids (Note: Refer to 49 CFR 178.502 and 178.503 for a complete explanation of markings.)**

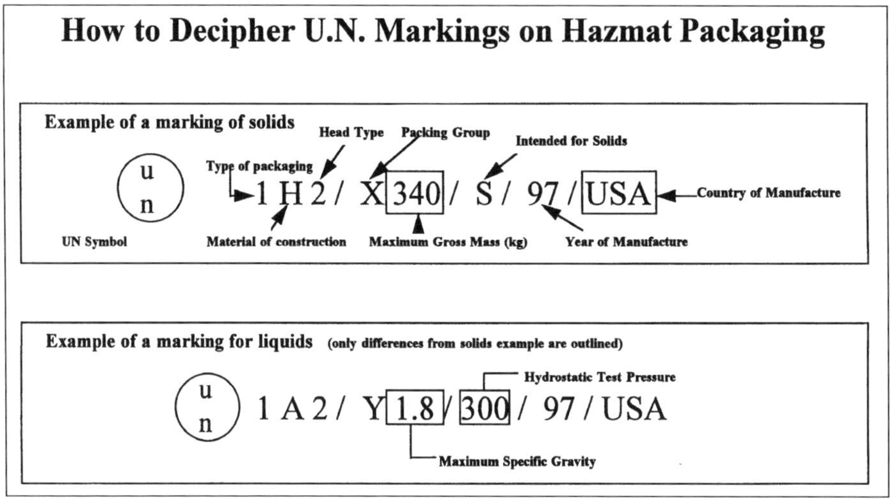

In the NAERG (see Exhibit 14), chemicals are listed in both alphabetical order (blue section) and UN/NA numerical order (yellow section) for reference by emergency responders. Each UN/NA number has an associated emergency response guide number (e.g., MEK, guide number 127; sulfuric acid, guide number 137) that details to a first responder at a hazardous materials incident the generic precautions to take with regard to emergency response activities.

On the manifest, the NAERG guide number for each hazardous waste in the shipment can be referenced in Section 15, Special Handling Instructions, or a copy of each applicable ERG guide page can be attached to the manifest prior to shipment.

The manifest also requires the generator to provide a 24-hour emergency response number of a person knowledgeable of the waste. This means that someone in your organization must be available to provide technical information in the event of an emergency.

Summary

Chapter 5 focused on the hazardous waste manifest—the requirements for completing the manifest and the distribution of each copy. The regulation has changed; no longer does each state require a copy. The chapter also featured instructions on how to complete a manifest. If a load is rejected at the TSDF,

Exhibit 14. **DOT North American Emergency Response Guidebook (NAERG)**

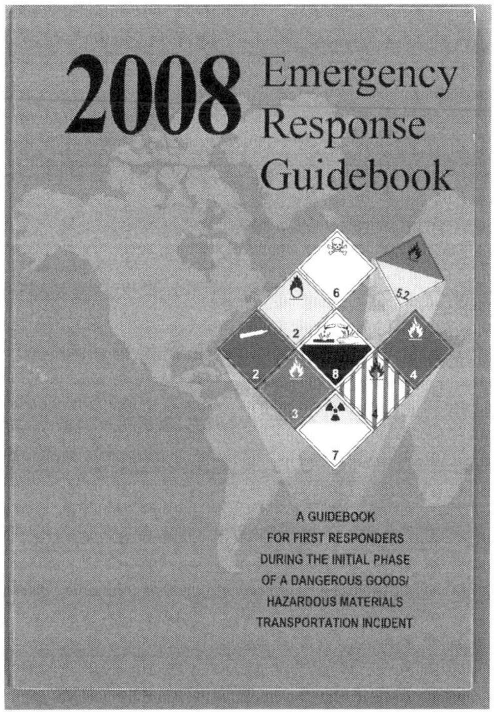

the load should return back to the generator. If a load has not been received within the required 45 days by the TSDF, the generator is required to notify their respective EPA or state environmental agency.

Chapter 6

Transportation of Wastes

Chapter 6 Learning Objectives

- Understand how to prepare a hazardous waste shipment for transportation following DOT requirements.
- Determine what planning is required to handle hazardous waste emergencies.
- Recognize the necessity of emergency response teams.

If the hazardous waste containers stored in an accumulation area are getting close to the 90-day regulatory deadline, a waste container pickup is necessary. As a rule of thumb, unless you usually ship the same waste streams with the same waste hauler, allow at least 30 days prior to needing a pick up to contact a hazardous waste transportation company. This should allow ample time to set up a convenient pickup date and create a contingency plan in case of an emergency (e.g., snowstorm, process upset). Remember, timing is everything with hazardous waste shipments.

In these times of competitiveness in the hazardous waste disposal market, some companies take inventory of what they need to ship out and get price quotes from two or three haulers. If there is time to do this, some money can be saved; however, from a liability standpoint to both the person signing the manifest and the hauling company, it's best to stick with one or two reputable waste haulers. This will be a benefit over time.

Another factor is to know where to send the waste. That's why auditing TSDF facilities and generating an internal approved facility list is important. Consistency in the transportation and disposal of hazardous waste is important.

First and foremost, ensure that the hazardous waste shipping containers to be used are DOT approved (HM-181 performance standards). Different hazardous materials have different requirements, but all containers must meet packaging specifications from the DOT, the primary regulating body for shipping hazardous waste. You are legally required to inspect every container prior to shipment to ensure that it meets these requirements.

To save time and money prior to a waste shipment, make sure that all the wastes have been profiled with the waste hauler and the TSDF. Generate an inventory list of the wastes being shipped. Ensure that the containers are labeled properly (DOT and hazardous waste labeling, if applicable) and that the writing on the labels is legible, that containers are in good condition (no large dents or rust on drums), and that container lids and bungs are tight. If a container needs to be overpacked (placed into a DOT-approved larger container) and the generating facility has the resources and the trained personnel to handle the job, it should do so.

Assign someone to oversee the shipping operation and have someone (such as a forklift driver) available when the containers are ready to be loaded onto the waste hauler; vehicle. On a hazardous waste shipment, time is money; better preparation means money saved.

Another time-consuming project is creating lab packs. Lab packing is the process of taking small chemical containers (usually lab quantities) and placing several containers into a large box or drum filled with absorbent material to protect the containers during shipment. Lab packs are necessary, for example, when the quality assurance or R&D lab on your facility has gone through its chemical inventory and found a fair number of containers that have past expiration dates or are close to the expiration date.

A good way to avoid trouble in this area is to send a memo informing everyone in the facility when a hazardous waste shipment is scheduled and ask each department if it has any materials that need to be disposed of. Set a reasonable time limit so that the necessary waste profile information [such as material safety data sheets (MSDSs)] can be prepared for the hauler so that the hauler brings the necessary materials for the job on the truck. Save time by accumulating all the lab pack material in one area for easy access (don't forget about chemical compatibility).

Another requirement for the generator is to offer the hazardous waste hauler DOT placards. Most experienced hazardous waste transporters have the placards available with the vehicle; however, it is required that the shipper offer the appropriate placards. Several sample placards are shown in the appendix.

If the NAERG guide number of the hazardous waste is referenced in in Section 15 of the manifest, the transporter is required to have a copy of the DOT North American Emergency Response Guidebook (NAERG) in its possession during transit. Prior to signing the manifest, ask the driver to see the guidebook. It is a good idea to keep a stock of the current version of the NAERG on hand.

Emergency Plans and Response

RCRA release-reporting requirements apply to generators, transporters, and TSDFs. RCRA release-reporting requirements vary according to whether one operates as a hazardous waste generator, a transporter, or a TSDF.

All RCRA hazardous wastes are also classified as hazardous substances under CERCLA (Superfund).

Materials such as methyl ethyl ketone (MEK) are listed in the DOT Hazardous Materials Table (49 CFR 172.101) in the reportable quantities (RQ) Appendix A. A facility shipping an amount of a hazardous waste that exceeds the reportable quantity should add the letters RQ to the proper shipping name on the manifest. Example: MEK has an RQ of 5000 pounds. If a facility is shipping a tank truck load of flammable waste with MEK being the prevalent constituent, the shipping name should read, "Waste flammable liquid, n.o.s. (contains methyl ethyl ketone) 3, UN 1193, PG III, RQ, D001." The RQ notation indicates that if there is a hazardous waste release of a certain amount, additional authorities must be notified.

Therefore, a release of a RCRA hazardous waste can be reportable under CERCLA and the Emergency Planning and Community Right-to-Know Act (EPCRA) and should follow their notification requirements. RCRA release-reporting requirements also involve notification of the EPA regional office and a written follow-up report within five days of the incident.

Hazardous waste transporters and TSDFs are also required to have a contingency plan that includes release-reporting steps and phone numbers of the agencies that should be contacted.

The only way to ensure compliance is to be prepared. Plan ahead:

- Record which hazardous substances at the facility may pose potential reportable releases.
- Investigate which federal statutes and regulations apply to the substances and practices at the facility.
- Know what information must be reported, who must be notified, and when the report must be made.
- Determine whether a written follow-up report is necessary, what information it should contain, how soon it should be filed, and where it should be sent.

Remember, even if a serious hazardous material spill has never occurred at a facility or on the highway, the possibility may still exist. Planning ahead and knowing whom to notify when an incident occurs is not only required by law but makes good sense.

Emergency Response Teams or Not?

An age-old question for most companies is, *Do we need to have an emergency response team on-site?* The answer depends on the company's culture and emergency response comfort level. Some industries, such as chemical, petrochemical, and nuclear, must have emergency response teams (ERTs) because of the nature of what they store, produce, or transport. The transportation industry (trucking, rail, maritime, air), also must include ERTs in their plans. Industries in which 29 CFR 1910.120, the Hazardous Waste Operation and Emergency Response (HAZWOPER) regulation applies (i.e., chemical, petrochemical) must develop and maintain on-site emergency response teams. For all other industries, the decision to have emergency response teams depends on the company's philosophy on crisis management.

The number of employees can help to determine whether an ERT can be accommodated (for example, the number of trained employees available on a shift or floor or in a department or plant at any given time). The OSHA HAZWOPER standard requires "buddy systems" for ERTs. For example, if two ERT members are preparing to combat a spill and don the required PPE, then two additional ERT members are required to be ready to respond in the event of an emergency. If a facility has fewer than four trained and competent operators available on a regular basis, this may not be an option.

ERTs expend a lot of resources of both time and money. The team must have the right equipment. Self-contained breathing apparatus (SCBA) and protective clothing such as total encapsulating (Level A) suits are expensive. The HAZWOPER regulation requires that the team has sufficient opportunities to practice their skills, which means time off the shop floor or overtime.

RCRA states that a facility's contingency plan (40 CFR 265.51–265.53) must be designed to minimize hazards to human health or the environment from fires, explosions, or any unplanned sudden or nonsudden release of hazardous waste or hazardous waste constituents to air, soil, or surface water. A facility must carry out this plan immediately whenever fires, explosion, or releases happen that could threaten people or the environment.

RCRA requires that a facility's contingency plan identify a primary emergency coordinator and alternates in ranking order and include their home addresses and phone numbers (office, cell, and home); that local authorities (e.g., police, fire department, hospital, local emergency planning committee (LEPC), state agencies) agree to the arrangements set forth in the plan; and that each of the aforementioned agencies that could provide emergency services maintain an updated copy of the plan.

RCRA further requires that a facility have spill response, decontamination, and fire extinguishing equipment available and a communications and alarm system (external and internal) and that equipment and systems are inventoried in the contingency plan. Evacuation plans and escape routes (primary and secondary) also must be included in the plan.

The decision to develop and maintain either a defensive or offensive on-site emergency response team to handle RCRA spills and releases is an individual company decision. The regulations allow facilities to rely on external emergency response agencies and contractors as long as the agencies and contractors maintain a current copy of the facility's contingency plan and are familiar with the facility.

There is a simple five-step emergency process that any organization can follow regardless of the severity of the incident:

1. **Evacuate the area**. As people leave the area, the contingency plan should have a mechanism in place to alert others.
2. **Observe while evacuating**. As people leave the area, they should try to observe and remember the nature of the accident, the type and amount of hazardous waste material involved, where the accident occurred, whether other people were involved, and the presence of fire or chemical reactions they may have seen. If an observation can be done safely, it will greatly help with the emergency response effort.
3. **Decontaminate affected workers**. If people involved have hazardous material anywhere on their body, they should know the locations of decontamination equipment (eyewash station/safety shower), first-aid kit, and emergency exits.
4. **Secure the area**. Block the area off right away to warn others of an emergency situation.
5. **Report hazardous waste emergencies**. Familiarize personnel with your company's emergency notification procedures.

Absorbents

If hazardous material clean-up options include using an absorbent material, there are a few issues to consider when choosing the material. Today, there are many different varieties of absorbents: clay, cellulose, cornhusks, sawdust, vermiculite, polyester, and polyurethane, which come in a variety of shapes: socks, mats, wipes, pads, booms, and pillows.

First, consider the chemicals the facility may need to clean up with the absorbent materials and conduct a compatibility study. The manufacturer should be able to provide technical information on which materials can be absorbed with their product. Some materials react with certain absorbents, so rather than have a clean-up scenario go sour, do some homework up front.

Second, check the compatibility of the absorbents with the TSDF or hazardous waste landfill to which the hazardous waste will be transported. Certain absorbent materials will not be accepted; remember, high BTU value is important to incinerators, while high BTU value and low residual ash are important to fuel blenders and kilns. Before making a purchase of the newest absorbent product on the market, check with a hazardous waste representative to inquire whether the absorbent will be accepted.

Third, hazardous waste is tallied by the pound. When you file biennial (every two year) reports to the state or EPA, the weight is registered. Some absorbent materials are lighter than others, thus producing a reduced weight of hazardous waste.

Review all three of the aforementioned items when selecting absorbent materials for a facility. Your facility's chemical list may require different types of absorbents. Make sure to train personnel on which absorbent is to be used with which materials.

Summary

In Chapter 6, we reviewed how to prepare a hazardous waste shipment for transportation under the DOT requirements. We learned that pickups should be scheduled to allow plenty of time so as not to exceed the 90- or 180-day time limits based on generator classification and that someone at a location should always be assigned to coordinate the shipment. We also discussed the emergency requirements for each participant in the cradle-to-grave process—generator, transporter, and TSDF. Finally, we reviewed the necessity for the generator to maintain and train an emergency response team based on the OSHA HAZWOPER regulation (29 CFR 1910.120).

Chapter 7

Personnel Training, Record Keeping, and Reporting

Chapter 7 Learning Objectives

- Determine who must be trained on the RCRA regulations.
- Understand the necessity for outlining RCRA job descriptions and job duties.
- Understand the record-keeping requirements for each type of document.

Personnel at a facility who engage in work that involves hazardous waste management must receive training by an RCRA-competent trainer. Training must be completed for a new hire or transferred employee within six months of the date of hire or transfer and annually thereafter. Each job must have a written job description detailing RCRA-related activities (40 CFR 265.16). (See Exhibit 15 for examples of required employee information.)

A competent RCRA trainer is a person who thoroughly understands the regulation and how it applies to the company or facility. The company decides how often the RCRA trainer requires refresher training. Some organizations hold annual train-the-trainer sessions for their site RCRA coordinators. This is wise because not only does it keep the site trainer refreshed on the regulation, it also allows the site RCRA coordinator to learn from others within the organization. Many private hazardous waste workshops are available annually around the country.

The regulation states that anyone who handles hazardous waste (including filling out RCRA-related paperwork such as manifests and waste profiles) must receive initial classroom or on-the-job training that describes the RCRA regulations and how they apply to the facility and the trainee's job responsibilities.

Exhibit 15. Required Employee Information That Must Be on File at All Hazardous Waste Generators

Examples of job responsibilities

1. **Operator A**

   Relief: Operator B

   - Load waste organics from closed loop condenser into tank truck for shipment to TSDF.

2. **Operator B**

   Relief: Operator A

   - Ensure that hazardous wastes are placed in approved containers that are compatible with the wastes to be containerized.
   - Transport containerized waste as needed to the hazardous waste accumulation area.
   - Be responsible for proper use of satellite accumulation areas.

3. **Shipping Supervisor**

   Relief: Production Coordinator

   - Oversee compliance with DOT hazardous material requirements as they apply to raw materials.
   - Supervise warehouse personnel in the performance of their hazardous waste duties.

4. **Environmental Coordinator**

   - Manage overall hazardous waste program.
   - Conduct on-the-job hazardous waste training for personnel handling hazardous waste.
   - Manifest hazardous waste to permitted TSDFs using permitted transporters.
   - Perform hazardous waste determination on waste streams, and, if necessary, arrange for profiling by off-site laboratories.
   - Act as alternative emergency coordinator to ensure that emergency response activities are properly handled.

5. **Emergency Coordinator**

   - Serve as primary emergency coordinator.
   - Ensure that emergency response activities are conducted in accordance with OSHA 1910.120.
   - Conduct hazard communication and emergency response training.

Training content must include how to properly handle hazardous waste and the facility's emergency procedures. Emergency procedures may be as simple as pulling an alarm or notifing someone and evacuating the area or as complex as mitigating the emergency. The most important thing for personnel to understand about emergencies is what they can and cannot do in an emergency scenario.

If applicable to a facility, emergency training should include

- procedures for using, inspecting, repairing, or replacing emergency or monitoring equipment;
- parameters for automatic waste feed cut-off systems;
- use of communications or alarm systems;
- response to fire or explosions;
- response to groundwater contamination incidents; and
- procedures for shutting down operations.

If personnel cannot complete this initial RCRA training during their orientation period, they must work under the supervision of a RCRA-trained individual until training is received, but not more than six months from date of hire or transfer.

Job Descriptions (40 CFR Part 265.16 (d)(1)(2))

Hazardous waste generators must maintain a current list of the job title for each position at the facility that is related to hazardous waste management and the name of the employee filling the position. A written job description for each position must include the requisite skill, education, or other qualifications and duties of facility personnel assigned to each position (see sample, Exhibit 16); a written description of the type of both introductory and continuing training that will be given to each person filling a position; and records documenting that the training or job experience required has been given to and completed by facility personnel. Additional RCRA job duties are listed in the appendix.

***Exhibit 16.* RCRA Job Titles and Qualifications (Sample)**

1. **Operator A**

   Employees filling position: John Smith
Pete Collins

   Relief: Operator B

   Qualifications:

   - Working knowledge of company chemical waste disposal procedures
   - High school education or equivalent
   - Hazard communications training/safety orientation

2. **Operator B**

   Employees filling position: Ron Peterson, Bill Atkins

   Relief: Operator A

   Qualifications:

   - Working knowledge of company chemical waste disposal procedures
   - High school education or equivalent
   - Hazard communications training/safety orientation

3. **Shipping Supervisor**

   Employee filling position: Albert Freeman

   Relief: Tom Johnson

   Qualifications:

   - Working knowledge of state and federal hazardous waste regulations
   - Working knowledge of company chemical waste disposal procedures

4. **Environmental Coordinator**

   Employee filling position: Sally Smith

   Qualifications:

   - Engineering or science degree from an accredited four-year college
   - Three years' experience in hazardous waste management
   - Detailed knowledge of company chemical waste disposal procedures
   - Completion of accredited hazardous waste management training

5. **Emergency Coordinator**

   Employee filling position: Larry Dyer

   Qualifications:

   - Trained hazardous materials specialist as specified in OSHA 1910.120
   - Detailed knowledge of site loss prevention and hazmat emergency system
   - Working knowledge of RCRA and CERCLA regulations
   - Eight years' work experience in safety, industrial hygiene and environmental field

Training records on current personnel must be kept until closure of the facility. Training records on former employees must be kept at least three years from the date the employee last worked at the facility. Training records may accompany a person transferred within the same company.

The key to RCRA training is ensuring that both the regulatory issues of RCRA and the facility-specific practices and procedures are covered in both orientation and annual training sessions. Demonstration of competency (tests) is the preferred method of gauging whether personnel understand the training provided.

Record Keeping and Reporting (40 CFR Part 262)

The record-keeping and reporting requirements for LQGs and SQGs provide the EPA and state agencies with a method of tracking the quantities of hazardous waste generated and the movement of such wastes. The generator regulations in 40 CFR Part 262 contain three primary record-keeping and reporting requirements:

- Biennial reporting
- Exception reporting
- Record retention.

Biennial Reporting (Every Two Years)

The biennial reporting requirements are intended to provide the EPA with reliable national data on hazardous waste management. In order to achieve this, LQGs who ships hazardous waste off-site to a TSDF or any size generator that treats, stores, or disposes of hazardous waste must submit a biennial report (EPA Forms 8700-13A and B) to the EPA regional administrator or the state by March 1 of each even-numbered year. The report details the generator's activities during the previous calendar year and includes

- EPA ID number, name, and address of the generator;
- EPA ID number and name of each transporter used throughout the year;
- EPA ID number, name, and address of each off-site TSDF and recycler to which the waste was sent during the year; and
- descriptions and quantities for each waste generated.

Most states will automatically send a facility a biennial report with the aforementioned information on the report. The RCRA coordinator must review the completed manifests and input the correct information.

The federal RCRA regulations do not require SQGs to file biennial reports.

Exception Reporting

The RCRA regulations ensure that the transportation of hazardous waste from its point of generation to its point of treatment, storage, or disposal is documented through the manifest system. This system requires the designated facility to return a signed and dated copy of the manifest to the generator in order to acknowledge the receipt of the waste. If the generator does not receive this paperwork, additional steps must be taken in order to locate the waste. As a result, LQGs that transport waste off-site but that do not receive a signed and dated copy of the manifest from the designated facility within 45 days from the date on which the initial transporter accepted the waste must submit an exception report to the EPA regional administrator. The exception report must describe efforts made to locate the waste and the results of those efforts.

SQGs that do not receive a signed and dated copy of a manifest from the designated facility within 60 days must send a copy of the original manifest to the EPA regional administrator with a note indicating that they have not received a return copy.

Record Retention

Generators must keep a copy of each biennial report and any exception reports for at least three years from the due date of the report. Generators are also required to keep copies of all manifests for three years or until a signed and dated copy of the manifest is received from the designated facility. The manifest received from the designated facility must be kept for at least three years from the date on which the hazardous waste was accepted by the initial transporter. Finally, records of waste analyses and determinations performed by the generator must be kept for at least three years from the date waste was last sent to an off-site TSDF. These retention periods may be extended automatically during the course of any unresolved enforcement action regarding the regulatory activity or as requested by the EPA administrator. As mentioned earlier, LDR forms must be kept for five years.

Summary

Chapter 7 walked through training requirements for the various generator classifications. Large-quantity generators (LQG) are required to develop and maintain job descriptions for all job classifications that have the potential to come in contact with hazardous waste. LQGs are also required to develop and maintain RCRA job titles and qualifications for individuals who will have responsibility in handling, shipping, completing paperwork, supervising, and coordinating hazardous waste emergencies. The last section of this chapter describes which RCRA documents (such as manifests, waste analyses, and LDR forms) must be retained and how long they must be retained.

Chapter 8

Selecting a Hazardous Waste Transporter and TSDF

Chapter 8 Learning Objectives

- Understand the objectives for selecting a hazardous waste transporter.
- Be able to select a hazardous waste TSDF.
- Understand Class I injection wells.
- Determine what should be included on a TSDF audit protocol.

When looking for hazardous waste transporters and TSDFs, the key word to remember is *liability*. A generating company has cradle-to-grave responsibility, whether the waste it generates is sitting on company property, in transit to the TSDF, parked at a transfer station in the middle of nowhere, or at the TSDF. Hazardous waste transporters provide guarantees, but the bottom line is that when an emergency occurs involving a facility's hazardous waste in the middle of the night, someone in the company must be ready to mitigate the situation, with either manpower or cash flow.

Transporters

Since hazardous waste transporters move regulated wastes on public roads and highways, rails, and waterways, they are regulated not only by RCRA but by the Department of Transportation (DOT) as well. To avoid regulatory discrepancies, the hazardous waste transporter regulations were developed jointly by the EPA and the DOT. Although these regulations are integrated, they are not located in the same part of the CFR. The DOT's Hazardous Materials Transportation Act (HMTA) regulations are found in 49 CFR parts 171–179, while the RCRA Subtitle C transporter requirements are located in 40 CFR Part 263.

Transporter regulations apply only to off-site transportation of hazardous waste. They do not apply to on-site transportation of hazardous waste within a facility's property or boundary.

Prior to selecting a hazardous waste transporter, facilities must do their homework. Salespeople will offer the sun and the moon; however, the bottom line is finding a transporter with dependability, reliability, regulatory knowledge, and a good transportation track record.

A facility must decide whether to select a TSDF with transportation capabilities or a waste broker that will subcontract transportation services. This determination should be based on the types and amounts of waste streams a company generates.

Once the list of possible hazardous waste transportation companies has been narrowed to two or three, contact the state environmental agency and ask if each transporter is licensed in the state and what its compliance record is (ask if there are any filed or pending notices of violations (NOVs)—in both EPA and DOT). Request each transporter's DOT number and go online at http// csa2010.fmcsa.dot.gov to the Comprehensive Safety Analysis 2010 (CSA 2010) System to review the company's driving safety and inspection records under seven Behavior Analysis Safety Improvement Categories (BASICS). Multi-site organizations can also check with other facilities in the organization that may have used the transporter to ask if the transporter is dependable. And finally, always ask the transporter for references and check the references.

Remember that under the Superfund (CERCLA) regulation (40 CFR Part 280.12), if a transporter spills hazardous waste in transit, the generating company will incur financial responsibility for the clean-up operation. Despite what a transporter may say, the generator is not released from any liability during the shipment phase.

Information to request from a transporter includes EPA identification number; DOT number; at what TSDF facilities it is permitted to transport waste streams; where waste will go when it leaves a facility (a transfer station? If so, where and for how long?); waste pickup scheduling lead time; regulatory background (Will it help you complete the waste profiles, manifests, and land ban forms?); what services it will provide on a facility's site (labeling, marking of containers, overpacking if necessary, lab-packing); training competencies of on-site technicians and drivers; emergency response capabilities; and, of course, pricing.

Transfer Facilities

Transporters accepting hazardous waste from a generator or another transporter may need to hold waste temporarily during the normal course of transportation. A transfer facility is defined as any transportation- related facility, such as a loading dock, parking area, storage area, or other similar area where shipments are held during the normal course of transportation. A transporter may hold waste at a transfer facility for up to ten days. Transfer facilities are usually used by transporters as a cost-saving measure to maximize trailer capacity on runs to different TSDF facilities.

Treatment, Storage, and Disposal Facilities (TSDFs)

Once a transporter has been chosen, a facility must decide where to send the waste. The final disposition of the waste will determine the type of TSDF to select. Incineration, fuel blending, recycling, landfilling, and Class I injection wells are the most common options. Pricing will play a big part, but once again, liability must be factored into the decision. Some TSDFs offer transportation services, and that tends to keep their prices competitive. This type of TSDF may be the best option for facilities with consistent waste streams. For facilities with a variety of waste streams that may need to go to different types of TSDFs, a waste broker may be the best bet.

Waste minimization or producing less waste is the best way to protect the environment and reduce liability risk. However, once waste is produced, the EPA recommends recycling as the next best method for managing it. Recycling offers the advantages of avoiding disposal costs and protecting the environment. But if mismanaged, recycling can release hazardous waste into the environment. RCRA, therefore, exclusively regulates wastes destined for recycling.

Recycling includes burning wastes for energy (40 CFR 261.6). This process must destroy at least 99.99 percent of the waste, which in turn minimizes a facility's risk of future liability that might have arisen if the waste were mismanaged. (The destruction efficiency is based on the type of incineration unit, temperature, and type of waste being burned.) Recycling usually produces a useful product from wastes, and it may preserve natural resources such as coal and fossil fuels used in the incineration process. Finally, depending on the material, recycling may be less expensive.

Incineration and fuel blending have become popular disposal or treatment options as the EPA continues to tighten regulations governing landfills and as future liabilities associated with buried wastes increase.

Land disposal of waste has the highest potential liability risk, and many consider it the method of last resort. The EPA has created the land disposal restriction (LDR) program, which prohibits placing untreated waste in land-based disposal or treatment units. Hazardous waste destined for land disposal must be pretreated to specified safe levels to reduce the risk to the environment.

Incineration and Fuel Blending

Incineration and fuel blending are both are forms of thermal destruction that break down hazardous components of waste to an efficiency of 99.99 percent.

The volume and weight of the waste are reduced to a fraction of their original state (referred to as ash content).

The big advantage of incineration and fuel blending is that many toxins are destroyed rather than merely contained. Risks to the environment and the generator are limited. While thermal destruction generally costs more than disposal in a landfill, increasing costs and limited availability of landfills are closing the gap.

There is a difference between fuel blending and incineration. Approximately 80 to 90 cement kilns operate in the United States, about 25 of which burn waste-derived fuel. This is known as fuel blending. Fuel is fed into the lower (hot) end of the kiln, where gas temperatures can reach 3500°F. As waste-derived fuels burn in the kiln, organic compounds are destroyed and inorganic compounds recombine with raw materials that are incorporated into gypsum used by ready-mix concrete makers.

Typically, fuel blending is the least expensive, most efficient form of incineration and has the least impact on the environment. Waste is converted into fuel with complete thermal destruction of organic waste. This waste-to-energy conversion preserves natural energy resources while destroying waste. Incineration, on the other hand, uses virgin fuels such as coal to fire an incinerator that burns at temperatures ranging from 1000°F to 2500°F. Because fuel blending uses wastes as energy, it's significantly less expensive than incineration.

Wastes that are high in British Thermal Unit (BTU) value and have low residual ash content are prime candidates for fuel blending. BTU values are important in the incineration process because the higher the BTU value, the more efficiently wastes will burn and convert to energy. For example, the most common use of absorbents is for daily maintenance to absorb oil leaks and spills; however, not all absorbents used for this purpose are good candidates for fuel blending. Clay pellets, a low grade of absorbent commonly used in industry, are unacceptable for fuel blending because of their 100 percent ash content remaining and low BTU value.

Wastes being sent for incineration and/or fuel-blending operations are used for their BTU value to keep that unit operational. So the higher the BTU value of a hazardous waste stream, the more likely it is that the incinerator or fuel-blending facility will accept the waste.

What Are Disposal Wells?

Disposal wells are vertical pipelines constructed of steel and concrete that are used to move liquids safely from the surface, through many rock layers, down to a deep disposal formation. Liquid wastes are safely and permanently managed by placing them into rocks that are naturally filled with salt water. Disposal occurs in layers thousands of feet deeper than any usable drinking water.

Class I wells provide multiple layers of waste isolation. At the surface, four layers of pipe and three layers of cement provide protection for all usable water. The entire well provides a minimum of two layers of pipe, one layer of cement, and a pressurized *annulus* (an open safety space between the two pipes) to isolate the waste. The installation is a well within a well. Literally hundreds of rock layers isolate waste placed in the disposal formation from drinking water and the surface environment. After disposal, the waste fluids are mixed with naturally generated brines (salt water) that have been trapped in the Earth's subsurface for millions of years.

The EPA provides oversight and closely monitors deep well disposal operations. In many cases, state governments have been granted primary authority to conduct an Underground Injection Control (UIC) program (40 CFR Part 148). In addition to strict construction requirements, annual testing and limitation of waste type, disposal volume, and injection well pressure are required to further promote protection of the accessible environment. Detailed records of wastes disposed and all site operations must be submitted to federal or state EPA regulators and made available to the public. Twenty-four-hour monitoring of operations is required to verify safe operation and permit compliance.

Class I Injection Well Technology

A wide variety of liquid wastes can be managed using deep injection well technology. Typically, these wastes are, low-concentration waste streams that are aqueous and comprised of more than 95 percent water. Often these liquids have already been recycled or pretreated and are not usable as a resource since they have little economic or energy value.

Some common dilute wastes now managed via deep well disposal include

- saltwater brines,
- dilute hydrocarbons and solvents,
- heavy metal solutions and agricultural wastewater,
- storm water and municipal waste,
- meat- and crop-processing liquids,
- inorganic solutions from manufacturing,
- industrial cleaning solutions, and
- groundwater clean-ups and landfill leachate.

The concentration of disposed materials in many of these injected wastes is similar that obtained by mixing a 2-gallon can of a product like gasoline into a large tank truck of fresh water (2 gallons/4000 gallons = 500 parts per million).

Disposal or injection wells are used throughout the world. In the United States, disposal wells associated with oil and gas production are regulated as Class II wells, while industrial injectors are regulated as Class I wells. In the oil and gas industry, disposal wells can be found where hydrocarbons are produced. Class I wells are used in areas where industrial needs as well as restrictive geologic and hydrologic standards can be met. A primary consideration in well location is the natural ability of underground formations to contain and isolate both natural and injected fluids. The geology in many areas of the United States, where a thick section of sedimentary rock is present, often meets these strict requirements. There are more than 200,000 disposal wells currently operating throughout the United States. Approximately 500 of these injectors are Class I wells.

Auditing the TSDF

Prior to sending waste to a TSDF facility, a company should develop some methodology of ensuring that its waste is in good hands for final disposal. Whether a facility is able to send someone to physically audit the TSDF facility or contracts with someone in the business of auditing TSDF facilities and selling reports is entirely up to the company. However, from a liability standpoint, don't just send waste to a facility and hope for the best.

Most TSDF facilities are not large and can usually be audited in a day or two. A lot of the regulatory paper review can be done prior to the audit. Most TSDF facilities schedule audits on certain days of the week and/or times of the month to maintain the least amount of interruption to their production schedule. A TSDF representative or waste broker representative can arrange an audit. With competition in the hazardous waste services field at its highest, most TSDF facilities have already developed a preliminary audit package for auditors to review. A waste management facility's evaluation document should include the following:

- **Facility History.** How many years in operation? what is EPA identification number?
- **Processing Capabilities.** What services are available on-site? What waste types are accepted and not accepted? What are the hours of operation? What is the weekly processing capacity?
- **Storage Practices and Capacities.** What is the permitted waste storage capacity? Cleanliness of truck unloading and drum storage areas: Are there signs of leakage or deterioration? Are there signs of a growing backlog of wastes? Air emission controls: Are the devices in good working condition?
- **Laboratory Capabilities and Waste Stream Qualifications.** Is an on-site laboratory available? What are its capabilities? Can facility prequalify waste streams?
- **Receiving and Loading Practices.** Does the facility test waste deliveries prior to unloading each truck? Is the generator or transporter notified of a rejection?
- **Waste Released by or Shipped from the Facility.** Are there any visible air emissions or detectable odors? How is storm water runoff from contained

areas controlled? Are any wastes or by-products shipped from the facility for off-site disposal or use as a fuel (tank bottoms, leachate, etc.)?

- **Regulatory Compliance.** Is the RCRA Part B permit in order? Do inspections from regulatory agencies (citations, corrective actions, compliance schedules) warrant notice? Is there an employee training program? Is the facility financially capable of handling environmental impairment liability? Is a security system in place? What is the description of the surrounding area (residential, industrial, wooded, etc.)?
- **Ground Water Monitoring and Leachate Control Programs.** Is the monitoring equipment working properly and are collections lagoons, ponds, etc. clean and free of debris? Are monitoring records available for inspection?
- **Other facilities owned or operated by company;**
- **Summary of concerns, comments, and observations.**

Exhibit 17 provides a checklist for a TSDF evaluation developed from many site visits. Although it may be hard to spare two people to conduct an audit, it is probably in a company's best interest to send someone who is familiar with both environmental and safety regulations to audit the TSDF. The length of time between audits is determined by how well the previous audit went; a three-year interval is a good rule of thumb. Sooner than three years is necessary if there is regulatory intervention of the TSDF or if poor management practice warrants it.

***Exhibit 17.* Key Audit Parameters for a Site Visit at a TSDF**

TSDF Visit Report

1. **Executive Summary**

   Name/Address/Contact(s)/Date visited

   Setting (industrial, residential, etc.)

   SERIOUS CONCERNS: past or present discharges to the ground, surface waters, or air that need to be addressed immediately, or evidence of past or present employee exposures

   MODERATE CONCERNS: similar to serious concerns, but do not need to be addressed immediately. Moderate concerns include technical design flaws that have the potential to allow significant releases or injuries to occur.

   MINOR CONCERNS: those with which the facility can probably function adequately but have the potential to result in injuries or impacts to the environment

2. **Facility Summary**

   Location/Services offered

   Operations and capacity/Generator use/Ownership

3. **Facility Setting**
   Facility history/Assessment of setting
4. **Operations**
   Waste receipt and analysis
   On-site handling and transfer
   Design and operations/Residue management
5. **Management Practices**
   Perceived management attitudes
   Personnel and qualifications/Training/Record keeping
   Preparedness and prevention
   Site security/Occupational safety
6. **Regulatory Compliance**
   Permit status
   Past and pending violations and consent orders
   Summary of regulatory compliance
7. **Information to Collect**
   Site location map
   Process flow diagram
   Company organization chart

Summary

Chapter 8 described what to look for when hiring a hazardous waste transporter. The chapter also describes what criteria to use when looking for a TSDF for your waste streams (both hazardous and non-hazardous). A sample TSDF audit outline to ensure that waste goes to a reputable establishment was included. A company should never chance sending its waste streams with just any transporter or to just any TSDF without doing its homework. The chapter introduces Class I injection or deep wells as an alternative disposal of liquid hazardous waste. Finally the chapter discusses the importance of auditing your TSDF and outlines what you as the client needs to audit.

Chapter 9

Waste Minimization

Chapter 9 Learning Objectives

- Explain the EPA definition of "waste minimization".
- Understand the minimum waste minimization program elements as outlined by the EPA.
- List the characteristics of companies with excellent records of pollution prevention.

Section 15 of the Uniform Hazardous Waste Manifest refers to the following statement in 40 CFR Part 262.27(a):

> *If I am a large quantity generator, I certify that I have a program in place to reduce the volume and toxicity of waste generated to the degree I have determined to be economically practicable and that I have selected the practicable method of treatment, storage or disposal currently available to me which minimizes the present and future threat to human health and the environment;* ***OR****, if I am a small quantity generator, I have made a good faith effort to minimize my waste generation and select the best waste management method that is available to me and that I can afford.*

And then the manifest is signed and dated.

The person signing might think, "It's just the manifest"; however, by signing on the dotted line, as the saying goes, he or she is committing the organization to establishing financially feasible waste minimization practices. Some states have waste minimization plans (pollution prevention, commonly known as P2, source reduction) in place that go beyond the federal RCRA program. The manifest is a legally binding document, and someone in the company signs off on that document.

The EPA definition of waste minimization is:

the reduction, to the extent feasible, of hazardous waste that is generated prior to treatment, storage, or disposal of the waste. It is defined as any source reduction or recycling activity that results in either

- *reduction of total volume of hazardous waste,*
- *reduction of toxicity of hazardous waste, or*
- *both.*

The EPA's minimum program elements for waste minimization include

Top Management Support—Part of site policies or charter; establishment of site goals for waste reduction; designation of a waste minimization coordinator/team; establishment of an incentive program for successful waste minimization ideas; and inclusion of waste minimization in an annual training program

Characterization of Waste Generation and Waste Management Costs—Two methods of reviewing waste generation are required: classifying the waste generated, and then identifying the costs associated with each waste stream and assigning the costs to the process or plant generating the waste. Classifying waste can be accomplished by any of these four methods or a combination of the methods:

- Volume of Waste—The hazardous waste generator's report is reviewed on an annual/biennial basis to determine the largest waste streams and where waste minimization will benefit the most.
- Hazard Classification—Wastes are classified according to toxicity. This classification is part of the basis of waste minimization efforts.
- Disposal Hierarchy—The final disposition of the wastes generated (see Exhibit 18).
- Disposal Vulnerability—The vulnerability of the disposal options for the waste streams generated. Vulnerability is based upon potential future liability associated with the disposal practice, disposal facility, or transportation of the material to the disposal facility.

Periodic Waste Minimization Assessments—Based on the classification of waste streams, a determination can be made as to which streams will benefit the site the most in a waste minimization effort. At this point the waste stream must be viewed from the point of generation rather than end-of-pipe.

Cost Allocation System—After using one of the above methods to classify waste streams, a cost allocation must be performed to determine the actual costs associated with generating waste and maintaining compliance rather than viewing waste generation as part of the overhead or price of doing business. The goal is to identify the specific process/product/facility that generates the waste and to use this information as an incentive to perform waste minimization or justification to not use a waste minimization technique.

Exhibit 18. **The Waste Management Hierarchy Contrasts Costs and Savings Using Different Approaches to Waste Management**

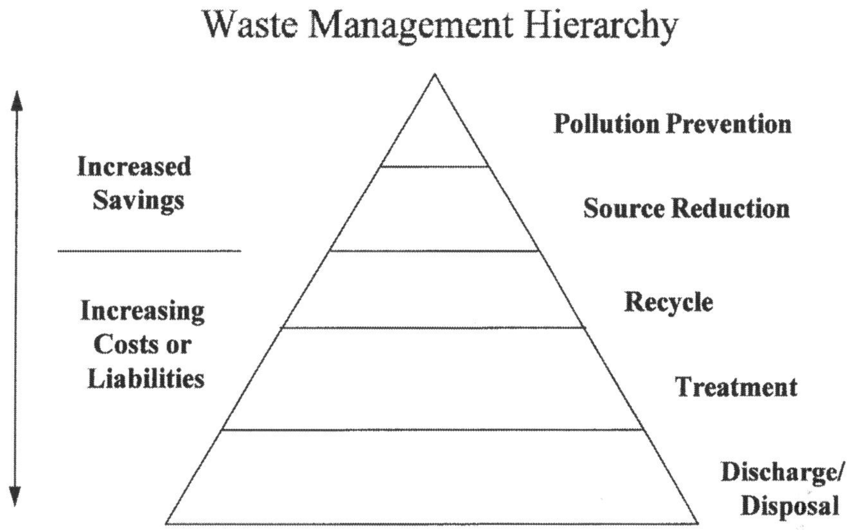

Technology Transfer—Advances in waste minimization, such as technology improvements, recycling alternatives, and material substitutions, should be distributed to others.

Program Implementation and Evaluation—The waste minimization plan is reviewed on a regular basis to determine general program effectiveness: Are waste minimization techniques being identified? Has implementation of waste minimization techniques been instituted in a timely fashion? Is training effective in involving site personnel? This review is conducted by the waste minimization team as part of its annual waste minimization review.

Characteristics of Companies with Excellence in Pollution Prevention

The following characteristics are repeatedly found in companies achieving excellence in pollution prevention:

- *Top-level support and commitment*—A clear commitment to pollution prevention by senior management is exhibited through policy, communications, and resources.
- *Proactive management*—Management is proactive rather than reactive. Actions are taken ahead of time in order to prevent pollution.
- *Interdisciplinary management*—A wide variety of people in the company are exposed to the ideas and techniques of pollution prevention.

Companies integrate the research, engineering, and production staffs into the pollution prevention programs. This increases understanding and commitment from the design staff to the sales force.

- *Corporate policies that integrate environmental issues*—pecific statements regarding environmental issues are incorporated into the company's goals and policies. These are often stated in a positive rather than a negative way (e.g., "Company policies should improve the environment," rather than "Company policies should not harm the environment").
- *Pollution prevention "Champion"*—Companies that excel in pollution prevention often have one or two key individuals—champions— who support and drive the programs. The champions believe in the program and push for its implementation.
- *High degree of employee awareness and training*—Thorough training of employees is essential in pollution prevention. Numerous examples exist of companies failing to prevent pollution simply because employees weren't trained properly.
- *Strong environmental auditing program*—Frequent assessments of pollution prevention include top-down, self-assessment, and third-party audits. These audits evaluate existing pollution prevention controls and measure pollution releases and reductions.

There are many methods by which a company can minimize waste: equipment or technology modifications; reformulation or redesign of products; substitution of raw materials; improvements in housekeeping and maintenance; and inventory control systems. Companies must choose the best method for their situation, obtain a commitment from management, and involve as many people as possible. Waste minimization, source reduction, pollution prevention—whatever phrase a company coins for reduction of hazardous and solid wastes will not only benefit the company's bottom line but also keep the environment healthy for generations to come.

Summary

Chapter 9 highlights the minimum EPA requirements for a pollution prevention and waste minimization program. The key element is having someone coordinate the efforts at a company to drive the initiative. Responsible companies recognize the benefit of waste minimization programs; they are not only good for the environment but cost effective as well. This cost effectiveness can give some companies a competitive advantage over their competitors because they spend less on waste and associated costs. Substitution of less-toxic materials or changing the process to generate more products and less waste are two simple examples of a company reducing its environmental output.

Appendix I

40 CFR Parts Impacting RCRA

Part	Title
Part 240	Guidelines for the thermal processing of solid wastes
Part 243	Guidelines for the storage and collection of residential, commercial, and institutional solid waste
Part 244	Solid waste management guidelines for beverage containers
Part 245	Promulgation of resource recovery facilities guidelines
Part 246	Source separation for materials recovery guidelines
Part 247	Comprehensive procurement guideline for products containing recovered materials
Part 254	Prior notice of citizen suits
Part 255	Identification of regions and agencies for solid waste management
Part 256	Guidelines for development and implementation of state solid waste management plans
Part 257	Criteria for classification of solid waste disposal facilities and practices
Part 258	Criteria for municipal solid waste landfills
Part 259	{Reserved}
Part 260	Hazardous waste management system: General
Part 261	Identification and listing of hazardous waste
Part 262	Standards applicable to generators of hazardous waste
Part 263	Standards applicable to tranporters of hazardous waste
Part 264	Standards for owners and operators of hazardous waste treatment, storage, and disposal facilities
Part 265	Interim status standards for owners and operators of hazardous waste treatment, storage, and disposal facilities
Part 266	Standards for the management of specific hazardous wastes and specific types of hazardous waste management facilities

Part	Title
Part 267	{Reserved}
Part 268	Land disposal restrictions
Part 270	EPA administered permit programs: The Hazardous Waste Permit Program
Part 271	Requirements for authorization of state hazardous waste programs
Part 272	Approved state hazardous waste management programs
Part 273	Standards for universal waste management
Part 279	Standards for the management of used oil
Part 280	Technical standards and corrective action requirements for owners and operators of underground storage tanks (USTs)
Part 281	Approval of state underground storage tank programs
Part 282	Approved underground storage programs
Parts 283	through 289 {Reserved}

29 CFR 1910 (OSHA) Regulations Impacting RCRA

Subpart	Title
H	Hazardous Materials
I	Personal Protective Equipment
K	Medical and First Aid
L	Fire Protection
Z	Toxic and Hazardous Substances

APPENDIX II

RCRA and Solid Waste Resources

Appendix II includes the URLs for only some Web sites; for consistency, recommend including the website addresses for all resources provided in this appendix and that each of the web URLs are formatted the same (currently some bold, some not; some listed as 'Home Page' while others listed as 'web site".

Hotlines

Emergency Planning Community Right-to-Know Hotline	
CERCLA (SARA, Title III) 8:30 a.m.–7:30 p.m. ET	1-800-424-9346
EPA Hotline (Region V)	1-800-621-8431
EPA RCRA, Superfund, Hazardous Waste Hotline Office of Solid Waste and Emergency Response 8:30 a.m.–7:30 p.m. ET	1-800-424-9346
National Response Center/Coast Guard Command (Report spills, chemical releases, radiological incidents)	1-800-424-8802
Solid Waste Assistance Program	1-800-677-9424

U.S. Federal Agencies and Offices

Environmental Protection Agency (EPA)
Ariel Rios Building
1200 Pennsylvania Avenue, NW
Washington, DC 20460 — 1-202-272-0167

EPA Web Site — http://www.epa.gov

EPA Region 1 (Boston, Massachusetts)	1-617-918-1111
EPA Region 2 (New York, New York)	1-212-637-3000
EPA Region 3 (Philadelphia, Pennsylvania)	1-215-814-5000
EPA Region 4 (Atlanta, Georgia)	1-404-562-9900
EPA Region 5 (Chicago, Illinois)	1-312-353-2000

EPA Region 6 (Dallas, Texas) 1-214-665-2200
EPA Region 7 (Kansas City, Kansas) 1-913-551-7003
EPA Region 8 (Denver, Colorado) 1-303-312-6312
EPA Region 9 (San Francisco, California) 1-415-947-8000
EPA Region 10 (Seattle, Washington) 1-206-553-1200

Public Information Center
401 M Street SW
Washington, DC 20460 1-202-260-7751
This EPA subsidiary maintains a wide selection of publications on major environmental topics. The materials distributed by PIC are nontechnical and have been prepared as sources of general environmental information for the public.

U.S. Government Printing Office
Superintendent of Documents
Post Office Box 371954
Pittsburgh, PA 15250-7954 1-202-512-1800
More than 15,000 books, pamphlets, posters, periodicals, subscription services, and other government publications are available for purchase from the superintendent of documents.

U.S. Department of Labor
Occupational Safety and Health Administration (OSHA)
200 Constitution Avenue, N.W.
Washington, DC 20210 1-202-693-2000
http://www.osha.gov

U.S. Department of Transportation
Hazardous Material Information Center 1-800-467-4922

Professional Societies/Associations and Private Agencies

Alliance of Hazardous Materials Professionals (AHMP)
Post Office Box 1216
Rockville, MD 20849 1-800-437-0137
Home Page http://www.achmm.org

Established in 1985, the AHMP is a nonprofit membership organization dedicated to fostering professional development through continuing education, peer group interaction, and the exchange of ideas and information relating to hazardous materials management.

Air & Waste Management Association (A&WMA)
1 Gateway Center, 3rd Floor
420 Fort Duquesne Boulevard
Pittsburgh, PA 15222 1-412-232-3444
Home Page http://www.awma.org
The Air and Waste Management Association is a nonprofit technical, scientific, and educational organization with more than 14,000 members in 65 countries. Founded in 1907, the association provides a forum where all viewpoints of an environmental issue (technical, scientific, economic, social, political, and risk assessment) receive consideration.

American Society of Safety Engineers (ASSE)
Environmental Practice Specialty
1800 East Oakton Street
Des Plaines, IL 60018-2187 1-847-699-2929
Home Page http://www.ASSE.org
The ASSE Environmental Practice targets members with formal education and experience in safety management, engineering, chemistry, health, physics, toxicology, management, and training.

Environmental Document Service
Ben Franklin Station
Post Office Box 7167
Washington, DC 20044 1-800-424-9068
A document service providing copies of environmental legislation, legal decisions and court filings, EPA rules, Superfund plans, and policy statements and directives.

Environmental Hazards Management Institute (EHMI)
10 Newmarket Road
Post Office Box 932
Durham, NH 03824 1-603-868-1496
Home page http://www.ehmi.org

A nonprofit environmental consulting corporation specializing in environmental and regulatory education and compliance assistance. Publishes the Household Hazardous Waste Wheel.

Government Institutes

4 Research Place, Suite 200

Rockville, MD 20850-1714 1-301-921-2300

A source for environmental and energy information. Selections include handbooks and manuals concerning RCRA, SARA, energy audits, and OSHA manuals.

National Recycling Coalition, Inc.

1220 L Street NW, Ste. 100-155 Washington, DC 20005 1-202-618-2107

Home page http://nrcrecycles.org

Washington, D.C.– based association whose members distill, blend recover, or recycle used chemicals and other hazardous and nonhazardous wastes.

APPENDIX III

Completed Hazardous Waste Manifest

Please print or type. (Form designed for use on elite (12-pitch) typewriter.) Form Approved. OMB No. 2050-0039

UNIFORM HAZARDOUS WASTE MANIFEST

1. Generator ID Number	2. Page 1 of	3. Emergency Response Phone	4. Manifest Tracking Number
ILD0123456789	1	(800) 555-1212	000695408 JJK

5. Generator's Name and Mailing Address — Generator's Site Address (if different than mailing address)

ABC Manufacturing, Inc.
3434 Bobcat Lane
Plainyille, Illinois 60000 (815) 555-0444

Generator's Phone:

6. Transporter 1 Company Name	U.S. EPA ID Number
Eagleway Transportation Services	ILD12345678
7. Transporter 2 Company Name	U.S. EPA ID Number

8. Designated Facility Name and Site Address — U.S. EPA ID Number: ILD234567890

Waste Stream Technologies
6263 Willington Lane
Wheretown, Illinois 60001 (815)555-2323

Facility's Phone:

9a. HM	9b. U.S. DOT Description (including Proper Shipping Name, Hazard Class, ID Number, and Packing Group (if any))	10. Containers No.	10. Containers Type	11. Total Quantity	12. Unit Wt./Vol.	13. Waste Codes
X	1. Waste Flammable Liquid, n.o.s. (contains Methyl Ethyl Ketone),3,UN1193 PG III	20	DM	1100	G	D001
X	2. Waste Corrosive Liquid, n.o.s. (contains sulfuric acid),8,UN1830,PG II,RQ, D002	18	DM	990	G	D002
X	3. Waste Paint Related Material, 3, UN1263 PG III, D001	4	DM	220	G	D001
	4. Used Oil - non-regulated	30	DM	1650	G	

14. Special Handling Instructions and Additional Information

1. NAERG #127   3. NAERG #128   If material cannot be delivered please return to generator.
2. NAERG #137

15. **GENERATOR'S/OFFEROR'S CERTIFICATION:** I hereby declare that the contents of this consignment are fully and accurately described above by the proper shipping name, and are classified, packaged, marked and labeled/placarded, and are in all respects in proper condition for transport according to applicable international and national governmental regulations. If export shipment and I am the Primary Exporter, I certify that the contents of this consignment conform to the terms of the attached EPA Acknowledgment of Consent.
I certify that the waste minimization statement identified in 40 CFR 262.27(a) (if I am a large quantity generator) or (b) (if I am a small quantity generator) is true.

Generator's/Offeror's Printed/Typed Name	Signature	Month	Day	Year
John M. Smith	John M. Smith	07	01	11

INT'L — 16. International Shipments ☐ Import to U.S. ☐ Export from U.S. Port of entry/exit: Transporter signature (for exports only): Date leaving U.S.:

TRANSPORTER — 17. Transporter Acknowledgment of Receipt of Materials

	Printed/Typed Name	Signature	Month	Day	Year
Transporter 1	Peter L. Calloway	Peter L. Calloway	07	01	11
Transporter 2					

DESIGNATED FACILITY —

18. Discrepancy

18a. Discrepancy Indication Space ☐ Quantity ☐ Type ☐ Residue ☐ Partial Rejection ☐ Full Rejection

Manifest Reference Number:

18b. Alternate Facility (or Generator) — U.S. EPA ID Number

Facility's Phone:

18c. Signature of Alternate Facility (or Generator) — Month Day Year

19. Hazardous Waste Report Management Method Codes (i.e., codes for hazardous waste treatment, disposal, and recycling systems)

1.	2.	3.	4.

20. Designated Facility Owner or Operator: Certification of receipt of hazardous materials covered by the manifest except as noted in Item 18a

Printed/Typed Name	Signature	Month	Day	Year
Susan A. Anthony	Susan A. Anthony	07	15	11

EPA Form 8700-22 (Rev. 3-05) Previous editions are obsolete.   DESIGNATED FACILITY TO DESTINATION STATE (IF REQUIRED)

Appendix IV

Example of Completed Land Disposal Restriction Form
(Please refer to accompanying Tables 1 and 2.)

LAND DISPOSAL RESTRICTION NOTIFICATION FORM 1

Page 1 of 1

Generator Name/Location ABC Manufacturing, Inc. Plainville, Illinois

EPA ID Number ILD0123456789 Manifest Number 000695408 JJK

Waste Analysis Available: X Yes ____ No X On file at facility

PROFILE #	RCRA NON-REGULATED Please check if waste stream is not regulated by RCRA	RCRA WASTE CODES (List all that apply)	SUBCATEGORY (See Table II and Select Key # if applicable)	TREATABILITY GROUP Please check the applicable treatability group		REGULATED CONSTITUENTS FOR F001, F002, F003, F004, F005	UNDERLYING HAZARDOUS CONSTITUENTS FOR D001*, D002, D003*, D004-D043
				Non-wastewater >1% TOC & >1% TSS	Wastewater	List all applicable constituents from key below	List all applicable constituents from Table 1
a	b	c	D	e	f	g	h
ABC123	X	D001	1	X		N/A	177

REGULATED CONSTITUENTS FOR F001, F002, F003, F004, F005, (for Column g)

1) Acetone
2) Benzene
3) N-Butyl Alcohol
4) Carbon Disulfide
5) Carbon Tetrachloride
6) Chlorobenzene
7) Cresols (o, m, or p isomers)
8) Cyclohexanone
9) o-Dichlorobenzene
10) Ethyl Acetate
11) Ethyl Benzene
12) Ethyl Ether
13) Isobutanol (Isobutyl alcohol)
14) Methanol
15) Methylene Chloride
16) Methyl Ethyl Ketone
17) Methyl Isobutyl Ketone
18) Nitrobenzene
19) Pyridine
20) Tetrachloroethylene
21) Toluene
22) 1,1,1 Trichloroethane
23) 1,1,2 Trichloroethane
24) 1,1,2 Trichloro 1,2,2 Trifluoroethane
25) Trichloroethylene
26) Trichlorofluoromethane
27) Xylene (Total)

I certify under penalty of law that the above information is accurate and true.

Signature *John M. Smith* Print Name John M. Smith Date 7/1/2011

Table 1 - UNIVERSAL TREATMENT STANDARDS
REGULATED CONSTITUENTS FOR D001*, D002, D012-D043, F039 (for column h)

#	Constituent
28)	Acenaphthylene
29)	Acenaphthene
30)	Acetone
31)	Acetonitrile
32)	Acetophenone
33)	2-Acetylaminofluorene
34)	Acrolein
35)	Acrylamide
36)	Acrylonitrile
37)	Aldicarb sulfone
38)	Aldrin
39)	4-Aminobiphenyl
40)	Aniline
41)	Anthracene
42)	Aramite
43)	alpha-BHC
44)	beta-BHC
45)	delta-BHC
46)	gamma-BHC
47)	Barban
48)	Bendiocarb
49)	Benomyl
50)	Benzene
51)	Benz(a)anthracene
52)	Benzal chloride
53)	Benzo(b)flouranthene (difficult to distinquish from benzo(k)fluoranthene)
54)	Benzo(k)fluoranthene (difficult to distinquish from benzo(b)fluoranthene)
55)	Benzo(g,h,i)perylene
56)	Benzo(a)pyrene
57)	Bromodichloromethane
58)	Bromomethane/Methyl bromide
59)	4-Bromophenyl phenyl ether
60)	n-Butyl alcohol
61)	Butylate
62)	Butyl benzyl phthalate
63)	2-sec-Butyl-4,6,-dinitrophenol/Dinoseb
64)	Carbaryl
65)	Carbenzadim
66)	Carbofuran
67)	Carbofuran phenol
68)	Carbon disulfide
69)	Carbon tetrachloride
70)	Carbonsulfan
71)	Chlordane (alpha and gamma isomers)
72)	p-Chloroaniline
73)	Chlorobenzene
74)	Chlorobenzilate
75)	2-Chloro-1,3-butadiene
76)	Chlorodibromomethane
77)	Chloroethane
78)	bis(2-Chloroethoxy)methane
79)	bis(2-Chloroethyl)ether
80)	Chloroform
81)	bis(2-Chloroisopropyl)ether
82)	p-Chloro-m-ceresol
83)	2-Chloroethyl vinyl ether
84)	Chloromethane/Methyl chloride
85)	2-Chloronaphthalene
86)	2-Chlorophenol
87)	3-Chloropropylene
88)	Chrysene
89)	o-Cresol
90)	m-Cresol (difficult to distinguish from p-cresol)
91)	p-Cresol (difficult to distinquish from m-cresol
92)	m-Cumenyl methylcarbamate
93)	Cyclohexanone
94)	o,p'-DDD
95)	p,p'-DDD
96)	o,p'-DDE
97)	p,p'-DDE
98)	o,p'-DDT
99)	p,p'-DDT
100)	Dibenz(a,h)anthracene
101)	Dibenz(a,e)pyrene
102)	1,2-Dibromo-3-chloropropane
103)	1,2-Dibromoethane/Ethylene dibromide
104)	Dibromomethane
105)	m-Dichlorobenzene
106)	o-Dichlorobenzene
107)	p-Dichlorobenzene
108)	Dichlorodifluoromethane
109)	1,1-Dichloroethane
110)	1,2-Dichloroethane
111)	1,1-Dichloroethylene
112)	trans-1,2-Dichloroethylene
113)	2,4-Dichlorophenol
114)	2,6-Dichlorophenol
115)	2,4-Dichlorophenoxyacetic acid/2,4-D
116)	1,2-Dichloropropane
117)	cis-1,3-Dichloropropylene
118)	trans-1,3-Dichloropropylene
119)	Dieldrin
120)	Diethyl phthalate
121)	p-Dimethylaminoazobenzene
122)	2-4-Dimethyl phenol
123)	Dimethyl Phthalate
124)	Di-n-butyl phthalate
125)	1,4-Dinitrobenzene
126)	4,6-Dinitro-o-cresol
127)	2,4-Dinitrophenol
128)	2,4-Dinitrotoluene
129)	2,6-Dinitrotoluene
130)	Di-n-octyl phthalate
131)	Di-n-propylnitrosamine
132)	1,4-Dioxane
133)	Diphenylamine (difficult to distinquish from diphenylnitrosamine)
134)	Diphenylnitrosamine (difficult to distinguish from diphenylamine)
135)	1,2-Diphenylhydrazine
136)	Disulfoton
137)	Dithiocarbamates (total)
138)	Endosulfan I
139)	Endosulfan II
140)	Endosulfan sulfate
141)	Endrin
142)	Endrin aldehyde
143)	EPTC[t]
144)	Ethyl acetate
145)	Ethyl benzene
146)	Ethyl cyanide/Propanenitrile
147)	Ethyl ether
148)	Ethyl methacrylate
149)	Ethylene oxide
150)	Famphur
151)	Fluoranthene
152)	Fluorene
153)	Formetanate hydrochloride
154)	Heptachlor
155)	Heptachlor epoxide
156)	Hexachlorobenzene
157)	Hexachlorobutadiene
158)	Hexachlorocyclopentadiene
159)	HxCDDs (All Hexachlorodibenzo-p-dioxins)
160)	HxCDFs (All Hexachlorodibenzofurans)
161)	Hexachloroethane
162)	Indeno (1,2,3-c,d) pyrene
163)	Iodomethane
164)	Isobutyl alcohol
165)	Isodrin
166)	Isosafrole
167)	Kepone
168)	Methacrylonitrile
169)	Methanol
170)	Methapyrilene
171)	Methicarb
172)	Methomyl
173)	Methoxychlor
174)	3-Methycholanthrene
175)	4,4-Methylene bis(2-chloroaniline)
176)	Methylene chloride
177)	Methyl ethyl ketone
178)	Methyl isobutyl ketone
179)	Methyl methacrylate
180)	Methyl methanesulfonate
181)	Methyl parathion
182)	Metolcarb
183)	Mexacarbate
184)	Molinate
185)	Naphthalene
186)	2-Naphthylamine
187)	o-Nitroaniline
188)	p-Nitroaniline
189)	Nitrobenzene
190)	5-Nitro-o-toluidine
191)	o-Nitrophenol
192)	p-Nitrophenol
193)	N-Nitrosodiethylamine
194)	N-Nitrosodimethylamine
195)	N-Nitroso-di-n-butylamine
196)	N-Nitrosomethylethylamine
197)	N-Nitrosomorpholine
198)	N-Nitrosopiperidine
199)	N-Nitosopyrrolidine
200)	Oxamyl
201)	Parathion
202)	Total PCBs (sum of all PCB isomers, or all Aroclors)
203)	Pebulate
204)	Pentachlorobenzene
205)	PeCDDs (All Pentachlorodibenzo-p-dioxins)
206)	PeCDFs (All Pentachlorodibenzofurans)
207)	Pentachloroethane
208)	Pentachloronitrobenzene
209)	Pentachlorophenol
210)	Phenacetin
211)	Phenanthrene
212)	Phenol
213)	Phorate
214)	Phthalic acid
215)	Phthalic anhydride
216)	Physostigmine
217)	Physostigmine salicylate
218)	Promecarb
219)	Pronamide
220)	Propham
221)	Propoxur
222)	Prosulfocarb
223)	Pyrene
224)	Pyridine
225)	Safrole
226)	Silvex/2,4,5-TP
227)	1,2,4,5-Tetrachlorobenzene
228)	TCDDs (All Tetrachlorodibenzo-p-dioxins)
229)	TCDFs (All Tetrachlorodibenzofurans)
230)	1,1,1,2-Tetrachloroethane
231)	1,1,2,2-Tetrachloroethane
232)	Tetrachloroethylene
233)	2,3,4,6-Tetrachlorophenol
234)	Thiodicarb
235)	Thiophanate-methyl
236)	Toluene
237)	Toxaphene
238)	Triallate
239)	Tribromomethane/Bromoform
240)	2,4,6-Tribromophenol
241)	1,2,4-Trichlorobenzene
242)	1,1,1-Trichloroethane
243)	1,1,2-Trichloroethane
244)	Trichloroethylene
245)	Trichloromonofluoromethane
246)	2,4,5-Trichlorophenol
247)	2,4,6-Trichlorophenol
248)	2,4,5-Trichlorophenoxyacetic acid/2,4,5-T
249)	1,2,3-Trichloropropane
250)	1,1,2-Trichloro-1,2,2-trifluoroethane
251)	Triethylamine
252)	tris-(2,3-Dibromopropyl) phosphate
253)	Vermolate
254)	Vinyl chloride
255)	Xylenes-mixed isomers (sum of o-,m-, and p-xylene concentrations)
256)	Antimony
257)	Arsenic
258)	Barium
259)	Beryllium
260)	Cadmium
261)	Chromium (Total)
262)	Cyanides (Total)
263)	Cyanides (Amenable)
264)	Fluoride
265)	Lead
266)	Mercury–Non-wastewater from Retort
267)	Mercury–All Others
268)	Nickel
269)	Selenium
270)	Silver
271)	Sulfide
272)	Thallium
272)	Vanadium
273)	Zinc

Table 2

The following waste codes have subcategories and the appropriate key number must be selected and placed in Column d on Form No. 1. Please refer to 40 CFR 268 for exact wording of subcategories.

WASTE CODES	KEY NUMBER	SUBCATEGORY
D001	1	High TOC ignitable liquids.
	2	Low TOC ignitable liquids managed in CWA/CWA-equivalent/Class 1 SDWA systems.
	3	Low TOC ignitable liquids managed in non-CWA/non-CWA-equivalent/non-Class 1 SDWA systems.
D002	4	Corrosive waste managed in non-CWA/non-CWA equivalent/non-Class 1 SDWA systems.
	5	Corrosive waste managed in CWA/CWA equivalent/Class 1 SDWA systems.
D003	6	Reactive Sulfides.
	7	Other Reactives.
	8	Water Reactive.
	9	Reactive Cyanide.
D006	10	Characteristic for Cd based on extraction procedure.
	11	Cadmium containing batteries.
D008	12	Characteristic for Pb based on extraction procedure.
	13	Lead Acid batteries.
D009	14	Low Mercury. (< 260 ppm total Hg)
	15	High Mercury. (≥ 260 ppm total Hg)
F003 F005	16	Wastes that contain only one or more of the following solvents: carbon disulfide, cyclohexanone, and/or methanol.
F005	17	Contains only 2-Nitropropane.
	18	Contains only 2-Ethoxyethanol.
F025	19	Light Ends.
	20	Spent Filters/Aids and Desiccants
K006	21	Anhydrous.
	22	Hydrated.
U151	23	Nonwastewaters that contain > 260mg/kg total mercury.
	24	All U151 (mercury) Wastewaters
K071	25	Nonwastewaters that are residues from RMERC.
	26	Nonwastewaters that are not residues from RMERC.
	27	All K071 Wastewaters.
P047	28	4,6-Dinitro-o-cresol.
	29	4,6-Dinitro-o-cresol salts.
P065	30	Nonwastewaters, not incinerator or RMERC residues.
	31	Nonwastewaters from RMERC w/less than 260 ppm Hg.
	32	Nonwastewaters from incinerator residues w/less than 260 ppm Hg.
	33	All P065 Wastewaters
P092	34	Nonwastewaters, not incinerator or RMERC residues.
	35	Nonwastewaters from RMERC w/less than 260 ppm Hg.
	36	Nonwastewaters from incinerator residues w/less than 260 ppm Hg.
	37	All P092 Wastewaters
U240	38	2,4-D (2,4-Dichlorophenoxyacetic acid).
	39	2,4-D (2,4-Dichlorophenoxyacetic acid) salts ad esters.

APPENDIX V

RCRA Site Identification Form

Instructions for completing EPA Form 8700-12 and an example of a first-time applicant.

Who Must Submit This Form

All sites required to submit any of the following must submit the RCRA Subtitle C Site Identification (Site ID) Form:

- Initial Notification of Regulated Waste Activity
- Subsequent Notification of Regulated Waste Activity
- First RCRA Hazardous Waste Part A Permit Application
- Revised RCRA Hazardous Waste Part A Permit Application
- Hazardous Waste Report
- Notification for eligible academic entities opting into or withdrawing from managing laboratory hazardous wastes pursuant to 40 CFR Part 262 Subpart K (if in an eligible state)
- Notification for facilities managing hazardous secondary material pursuant to 40 CFR 260.42 (if in an eligible state).

Some states have requirements in addition to, or that are different from, the federal requirements. To obtain the appropriate forms or ask questions, refer to a list of contacts at http://www.epa.gov/osw/inforesources/data/form8700/contact.pdf. The list will tell you whether the federal form or a state form is used, whom to contact, and where to mail the completed form.

Purpose of This Form

The Site ID Form provides site-specific information about a facility and is used to obtain an EPA ID number. The Site ID form also provides updated information for items that have changed at your site and verifies the information for those items that remain unchanged.

How to Fill Out This Form

Complete the following Site ID Form items as applicable to your facility:

- Item 1 – reason for submitting the form
- Item 2 – the site's EPA ID number

- Item 3 – the name of the site
- Item 4 – the physical location of the site
- Item 5 – the land type of the site
- Item 6 – the North American Industry Classification System (NAICS) code(s) for the site
- Item 7 – the mailing address for the site
- Item 8 – name, title, address, phone number, fax, and e-mail of a contact person at the site
- Item 9 – name, address, and phone number of the legal owner(s) and name of the operator(s) of the site
- Item 10 – the site's regulated waste activities (enter all that apply)
- Item 11 – description of the hazardous waste produced at the site
- Item 12 – the site's hazardous secondary material activity, if any are managed
- Item 13 – additional comments on Items 1–12
- Item 14 – certification that the information provided throughout the form is truthful, accurate, and complete
- Addendum to the Site Identification Form—notification of hazardous secondary material activity

Type or print, in black ink, all items except the signature box in Item 14. Enter the site's EPA ID number in the top left-hand corner on all pages of the form; for an Initial Notification for this site, leave the EPA ID number blank.

Reason for Submittal: Place an *X* in the appropriate box(es) to indicate whether this form is the Initial Notification (i.e., this is the first time the site is submitting site identification information / to obtain an EPA ID number for this location); a Subsequent Notification (to update site identification information); a component of a First or a Revised Hazardous Waste Part A Permit Application; or a component of the Hazardous Waste Report.

To provide an Initial Notification (first time submitting site identification information / to obtain an EPA ID number for this location).

- If the waste activity is regulated under Subtitle C of the Resource Conservation and Recovery Act (RCRA) and the rules promulgated pursuant to the Act (specifically 40CFR Parts 260-299), this form must be submitted to notify the appropriate state or EPA regional office of the site's regulated waste activities and obtain an EPA ID number.

- If the site is an eligible academic entity opting into 40 CFR 262 Subpart K for managing laboratory hazardous wastes **AND** it has never before submitted site identification information, this form must be submitted to notify the appropriate state or EPA regional office of the site's activities.

  Note: Check with the state to determine whether the site is eligible to manage laboratory hazardous waste pursuant to 40 CFR Part 262 Subpart K in order to notify.
- If the site will begin managing hazardous secondary material under 40 CFR 261.2(a)(2)(ii),40 CFR 261.4(a)(23), (24), or (25) **AND** your site identification information has never before been submitted, this form must be submitted, pursuant to 40 CFR 260.42, to notify the appropriate state or EPA regional office of the site's activities.

  Note: Check with the state to determine whether the site is eligible to manage hazardous secondary material under these exclusions in order to notify.

To provide a Subsequent Notification (to update site identification information for this location).

- Use this form to submit a subsequent notification if the site already has an EPA ID number and information must be changed (e.g., generator status, new site contact person, new owner, new mailing address, new regulated waste activity).
- If site identification information has previously been submitted and the site is an eligible academic entity opting into or withdrawing from 40 CFR Part 262 Subpart K for managing laboratory hazardous wastes, this form must be used.

  Note: Check with the state to determine eligibility to manage laboratory hazardous waste pursuant to 40 CFR Part 262 Subpart K in order to notify.
- If site identification information has previously been sumitted the site is notifying (or renotifying) that it will begin managing, is managing, or has stopped managing hazardous secondary material under 40 CFR 261.2(a)(2)(ii), 40 CFR 261.4(a)(23), (24), or (25), this form must be submitted pursuant to 40 CFR 260.42 to notify the appropriate state or EPA regional office of the site's activities.

  Note: Check with the state to determine whether the site is eligible to manage hazardous secondary material under these exclusions in order to notify.

OMB# 2050-0024; Expires 11/30/2011

SEND COMPLETED FORM TO: The Appropriate State or Regional Office.	United States Environmental Protection Agency **RCRA SUBTITLE C SITE IDENTIFICATION FORM**
1. Reason for Submittal MARK ALL BOX(ES) THAT APPLY	**Reason for Submittal:** ☒ To provide an Initial Notification (first time submitting site identification information / to obtain an EPA ID number for this location) ☐ To provide a Subsequent Notification (to update site identification information for this location) ☐ As a component of a First RCRA Hazardous Waste Part A Permit Application ☐ As a component of a Revised RCRA Hazardous Waste Part A Permit Application (Amendment #________) ☐ As a component of the Hazardous Waste Report (If marked, see sub-bullet below) ☐ Site was a TSD facility and/or generator of ≥1,000 kg of hazardous waste, >1 kg of acute hazardous waste, or >100 kg of acute hazardous waste spill cleanup in one or more months of the report year (or State equivalent LQG regulations)
2. Site EPA ID Number	**EPA ID Number**
3. Site Name	**Name:** ABC Manufacturing
4. Site Location Information	**Street Address:** 3434 Bobcat Lane **City, Town, or Village:** Anywhere **County:** Any **State:** Any State **Country:** USA **Zip Code:** 12345
5. Site Land Type	☒ Private ☐ County ☐ District ☐ Federal ☐ Tribal ☐ Municipal ☐ State ☐ Other
6. NAICS Code(s) for the Site (at least 5-digit codes)	A. 1 1 2 3 4 5 6 C. B. D.
7. Site Mailing Address	**Street or P.O. Box:** Same as #4 **City, Town, or Village:** **State:** **Country:** **Zip Code:**
8. Site Contact Person	**First Name:** John **MI:** M. **Last:** Smith **Title:** Environmental Coordinator **Street or P.O. Box:** Same as #4 **City, Town or Village:** **State:** **Country:** **Zip Code:** **Email:** JMSMITH@ABC.COM **Phone:** (815) 555-1212 **Ext.:** 123 **Fax:** (815) 555-1213
9. Legal Owner and Operator of the Site	**A. Name of Site's Legal Owner:** Mega Corporation **Date Became Owner:** 01/01/2010 **Owner Type:** ☒ Private ☐ County ☐ District ☐ Federal ☐ Tribal ☐ Municipal ☐ State ☐ Other **Street or P.O. Box:** 1234 Millwright Road **City, Town, or Village:** Big City **Phone:** (212) 555-1313 **State:** Any State **Country:** USA **Zip Code:** 01234 **B. Name of Site's Operator:** ABC Manufacturing **Date Became Operator:** 01/01/2010 **Operator Type:** ☒ Private ☐ County ☐ District ☐ Federal ☐ Tribal ☐ Municipal ☐ State ☐ Other

EPA Form 8700-12, 8700-13 A/B, 8700-23 (Revised 11/2009) Page 1 of 4

EPA ID Number

OMB#: 2050-0024; Expires 11/30/2011

10. Type of Regulated Waste Activity (at your site)
Mark "Yes" or "No" for all <u>current</u> activities (as of the date submitting the form); complete any additional boxes as instructed.

A. Hazardous Waste Activities; Complete all parts 1-7.

Y ☒ N ☐ **1. Generator of Hazardous Waste**
If "Yes", mark only one of the following – a, b, or c.

☒ a. LQG: Generates, in any calendar month, 1,000 kg/mo (2,200 lbs./mo.) or more of hazardous waste; **or** Generates, in any calendar month, or accumulates at any time, more than 1 kg/mo (2.2 lbs./mo) of acute hazardous waste; **or** Generates, in any calendar month, **or** accumulates at any time, more than 100 kg/mo (220 lbs./mo) of acute hazardous spill cleanup material.

☐ b. SQG: 100 to 1,000 kg/mo (220 – 2,200 lbs./mo) of non-acute hazardous waste.

☐ c. CESQG: Less than 100 kg/mo (220 lbs./mo) of non-acute hazardous waste.

If "Yes" above, indicate other generator activities.

Y ☐ N ☒ d. Short-Term Generator (generate from a short-term or one-time event and not from on-going processes). If "Yes", provide an explanation in the Comments section.

Y ☐ N ☒ e. United States Importer of Hazardous Waste

Y ☐ N ☒ f. Mixed Waste (hazardous and radioactive) Generator

Y ☐ N ☒ **2. Transporter of Hazardous Waste**
If "Yes", mark all that apply.

☐ a. Transporter

☐ b. Transfer Facility (at your site)

Y ☐ N ☒ **3. Treater, Storer, or Disposer of Hazardous Waste** Note: A hazardous waste permit is required for these activities.

Y ☐ N ☒ **4. Recycler of Hazardous Waste**

Y ☐ N ☒ **5. Exempt Boiler and/or Industrial Furnace**
If "Yes", mark all that apply.

☐ a. Small Quantity On-site Burner Exemption

☐ b. Smelting, Melting, and Refining Furnace Exemption

Y ☐ N ☒ **6. Underground Injection Control**

Y ☐ N ☒ **7. Receives Hazardous Waste from Off-site**

B. Universal Waste Activities; Complete all parts 1-2.

Y ☒ N ☐ **1. Large Quantity Handler of Universal Waste (you accumulate 5,000 kg or more) [refer to your State regulations to determine what is regulated]. Indicate types of universal waste managed at your site. If "Yes", mark all that apply.**

a. Batteries ☒
b. Pesticides ☐
c. Mercury containing equipment ☒
d. Lamps ☒
e. Other (specify) ____________ ☐
f. Other (specify) ____________ ☐
g. Other (specify) ____________ ☐

Y ☐ N ☒ **2. Destination Facility for Universal Waste**
Note: A hazardous waste permit may be required for this activity.

C. Used Oil Activities; Complete all parts 1-4.

Y ☐ N ☒ **1. Used Oil Transporter**
If "Yes", mark all that apply.

☐ a. Transporter

☐ b. Transfer Facility (at your site)

Y ☐ N ☒ **2. Used Oil Processor and/or Re-refiner**
If "Yes", mark all that apply.

☐ a. Processor

☐ b. Re-refiner

Y ☐ N ☒ **3. Off-Specification Used Oil Burner**

Y ☐ N ☒ **4. Used Oil Fuel Marketer**
If "Yes", mark all that apply.

☐ a. Marketer Who Directs Shipment of Off-Specification Used Oil to Off-Specification Used Oil Burner

☐ b. Marketer Who First Claims the Used Oil Meets the Specifications

EPA ID Number |_|_|_|_|_|_|_|_|_|_|_|_|_|_|_|_| OMB#: 2050-0024; Expires 11/30/2011

D. Eligible Academic Entities with Laboratories—Notification for opting into or withdrawing from managing laboratory hazardous wastes pursuant to 40 CFR Part 262 Subpart K

- You must check with your State to determine if you are eligible to manage laboratory hazardous wastes pursuant to 40 CFR Part 262 Subpart K

☐ 1. Opting into or currently operating under 40 CFR Part 262 Subpart K for the management of hazardous wastes in laboratories
See the item-by-item instructions for definitions of types of eligible academic entities. Mark all that apply:

☐ a. College or University

☐ b. Teaching Hospital that is owned by or has a formal written affiliation agreement with a college or university

☐ c. Non-profit Institute that is owned by or has a formal written affiliation agreement with a college or university

☐ 2. Withdrawing from 40 CFR Part 262 Subpart K for the management of hazardous wastes in laboratories

11. Description of Hazardous Waste

A. Waste Codes for Federally Regulated Hazardous Wastes. Please list the waste codes of the Federal hazardous wastes handled at your site. List them in the order they are presented in the regulations (e.g., D001, D003, F007, U112). Use an additional page if more spaces are needed.

D001	D002	F001				

B. Waste Codes for State-Regulated (i.e., non-Federal) Hazardous Wastes. Please list the waste codes of the State-Regulated hazardous wastes handled at your site. List them in the order they are presented in the regulations. Use an additional page if more spaces are needed.

EPA Form 8700-12, 8700-13 A/B, 8700-23 (Revised 11/2009) Page 3 of 4

EPA ID Number |_|_|_|_|_|_|_|_|_|_|_|_| OMB#: 2050-0024; Expires 11/30/2011

12. Notification of Hazardous Secondary Material (HSM) Activity

Y ☐ N ☒ Are you notifying under 40 CFR 260.42 that you will begin managing, are managing, or will stop managing hazardous secondary material under 40 CFR 261.2(a)(2)(ii), 40 CFR 261.4(a)(23), (24), or (25)?

If "Yes", you must fill out the Addendum to the Site Identification Form: Notification for Managing Hazardous Secondary Material.

13. Comments

14. Certification. I certify under penalty of law that this document and all attachments were prepared under my direction or supervision in accordance with a system designed to assure that qualified personnel properly gather and evaluate the information submitted. Based on my inquiry of the person or persons who manage the system, or those persons directly responsible for gathering the information, the information submitted is, to the best of my knowledge and belief, true, accurate, and complete. I am aware that there are significant penalties for submitting false information, including the possibility of fines and imprisonment for knowing violations. For the RCRA Hazardous Waste Part A Permit Application, all owner(s) and operator(s) must sign (see 40 CFR 270.10(b) and 270.11).

Signature of legal owner, operator, or an authorized representative	Name and Official Title (type or print)	Date Signed (mm/dd/yyyy)
Robert Jones	Robert Jones, President	01/01/2010

EPA Form 8700-12, 8700-13 A/B, 8700-23 (Revised 11/2009) Page 4 of 4

Appendix VI

Example of a Residential Household Chemical Waste Collection Program

(Courtesy of Solid Waste Agency of Lake County, Illinois.)

Household Chemical Waste Collections

Residents can dispose of their Household Chemical Waste (HCW) at SWALCO's (Solid Waste Agency of Lake County, IL) Year-Round Public Drop-off Events at the Gurnee facility or at one of the Mobile Events held at locations around Lake County.

Year-Round Public Drop-off Events:

Residents can dispose of their HCW at SWALCO's Year-Round Public Drop-Off Collection Events held at 1311 N. Estes Street in Gurnee.

Year-Round Collection Events are held twice a month, the **second Saturday and the fourth Monday,** at SWALCO's Gurnee facility **by appointment only**. Residents can schedule their appointment online or call the SWALCO office.

www.swalco.org
847-336-9340

SWALCO

SOLID WASTE AGENCY OF LAKE COUNTY, IL

Mobile Events:

SWALCO's Mobile Events are held at locations throughout Lake County from April through November (no appointment necessary). Check the website for upcoming events.

All HCW events sponsored by SWALCO are free for Illinois residents. Sign up online for SWALCO's elist for general announcements and information on SWALCO's programs, services and collection events.

Accepted Materials Include:

- Fertilizers/Herbicides/Pesticides
- Fluorescent Light Bulbs/CFLs
- Household Batteries (Rechargeable & Metal Hydride)
- Household Cleaners
- Oil-Based Paints/Paint Remover/ Varnishes
- Old Gasoline and Motor Oil
- Pool Chemicals
- Unwanted Prescription and Over-the Counter Medicines & Supplements (people and pet)

Not Accepted includes: Latex Paints, Alkaline Batteries, DEA Controlled Substances & Household Electronics.

APPENDIX VII

Required and Optional Labels for Shipping Containers with RCRA and Non-RCRA Waste

Nonhazardous Waste

For example, liquid styrene is spilled on ground and cleaned up with clay absorbent. The now-solid material is lab tested and deemed nonhazardous, but the containers are still shipped out as waste and disposed of properly (incineration).

Non-RCRA Regulated Waste

This label can be used for waste streams that do not meet the federal RCRA waste codes but meet state-specific waste codes (i.e., used oil in California and Massachusetts).

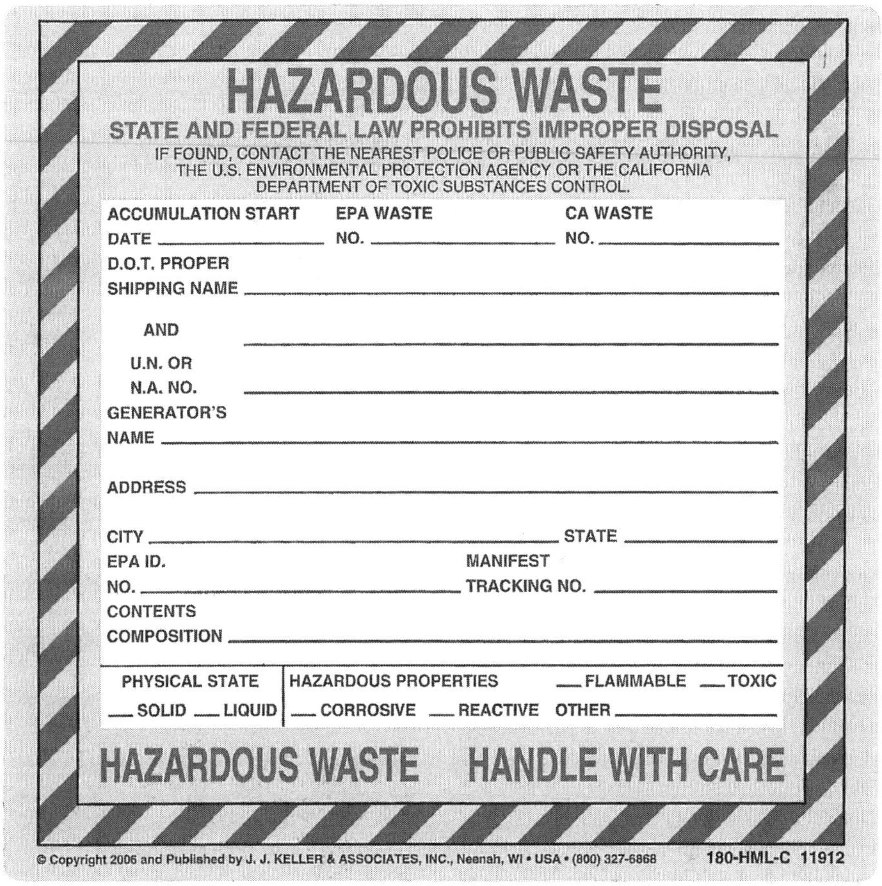
HAZARDOUS WASTE

STATE AND FEDERAL LAW PROHIBITS IMPROPER DISPOSAL

IF FOUND, CONTACT THE NEAREST POLICE OR PUBLIC SAFETY AUTHORITY, THE U.S. ENVIRONMENTAL PROTECTION AGENCY OR THE CALIFORNIA DEPARTMENT OF TOXIC SUBSTANCES CONTROL.

ACCUMULATION START DATE ______ EPA WASTE NO. ______ CA WASTE NO. ______

D.O.T. PROPER SHIPPING NAME ______

AND ______

U.N. OR N.A. NO. ______

GENERATOR'S NAME ______

ADDRESS ______

CITY ______ STATE ______

EPA ID. NO. ______ MANIFEST TRACKING NO. ______

CONTENTS COMPOSITION ______

PHYSICAL STATE ___ SOLID ___ LIQUID

HAZARDOUS PROPERTIES ___ FLAMMABLE ___ TOXIC ___ CORROSIVE ___ REACTIVE OTHER ______

HAZARDOUS WASTE HANDLE WITH CARE

© Copyright 2006 and Published by J. J. KELLER & ASSOCIATES, INC., Neenah, WI • USA • (800) 327-6868 180-HML-C 11912

State of California State-Specific Hazardous Waste Label

This label is required for waste created and shipped within the state of California (New Jersey also has a state-specific label—not shown).

Triple Rinse

Label is optional but used for containers that have been triple rinsed and determined to be RCRA clean.

UNIVERSAL WASTE

FEDERAL LAW PROHIBITS IMPROPER DISPOSAL

THE FOLLOWING MATERIALS ARE REGULATED AS A UNIVERSAL WASTE IN ACCORDANCE WITH 40 CFR PART 273.

- ☐ UNIVERSAL WASTE - BATTERY(IES)
- ☐ UNIVERSAL WASTE - MERCURY THERMOSTAT(S)
- ☐ UNIVERSAL WASTE - MERCURY CONTAINING EQUIPMENT
- ☐ UNIVERSAL WASTE - PESTICIDE(S)
- ☐ UNIVERSAL WASTE - LAMP(S)

ACCUMULATION START DATE: ____________________

____

____

____

D.O.T. PROPER SHIPPING NAME AND UN OR NA NO. WITH PREFIX (REQUIRED DURING TRANSPORT, WHEN MATERIAL IS ALSO REGULATED BY 49CFR PARTS 172-180)

HANDLE WITH CARE!

Style UW05VT © 2005 LABELMASTER® (800) 621-5808 www.labelmaster.com

Universal Waste

The universal waste label can be used for any of the following: batteries (lead acid); mercury thermostats or mercury-containing equipment (such as lamps and laboratory equipment); pesticides; and lamps (unbroken fluorescent bulbs).

Used Batteries

This label can be used for locations that collect batteries for recycling such as nickel-cadmium (ni-cad) batteries from two-way radios or alkaline batteries from many devices.

Used Oil

Used oil label is for containers with non-regulated oil usually sent for recycling.

Glossary

acutely hazardous waste—commercial chemical products and manufacturing intermediates having the generic names listed in 40 CFR 261.33; off-specification commercial chemical products and manufacturing chemical intermediates which, if they met specifications, would have generic names listed; and any residue or contaminated soil, water, or other debris resulting from clean-up of a spill of any of these substances. Also, even after containers of acutely hazardous waste are emptied, the containers themselves become hazardous waste.

accumulated speculatively—storage of a material in lieu of expeditious recycling. Materials are usually accumulated speculatively if the waste being stored has no viable market or if a facility cannot demonstrate that at least 75 percent of the material has been recycled in a calendar year.

best demonstrated available technology (BDAT)—the technology the EPA establishes for a land-banned hazardous waste to reduce the overall toxicity or mobility of toxic constituents in the waste. BDAT must be applied to such a waste prior to land disposal unless one can successfully demonstrate the validity of an equivalent treatment method.

bonding—connecting two objects by means of a clamp and bare wire. Its purpose is to equalize the electrical potential between the objects to prevent a static discharge when transferring a flammable liquid from one container to another. The conductive path is provided by clamps that make contact with the charged object and a low-resistance flexible cable, which allows the charge to equalize.

by-product—a material that is not one of the primary products of a production process and is not solely or separately produced by the production process.

CAS Number (CAS Registration Number)—an assigned number used to identify a chemical. CAS stands for Chemical Abstracts Service, an organization that indexes information published by the American Chemical Society and provides index guides with which information about particular substances may be located in the abstracts. The CAS number is a concise and unique means of material identification.

chain of custody—documentation of the who, where, how, and why of a sample; includes appropriately labeling the sample and identifies who picked it up, where it was acquired, how it was preserved, and who received it.

characteristic(s)—federal EPA-defined criteria by which a waste may be considered to have any of four characteristics that would cause it to be defined as hazardous. Characteristics are assigned "D" numbers for waste codes.

CERCLA—Comprehensive Environmental Response, Compensation and Liability Act of 1980, as amended (40 CFR 280.12). Also known as Superfund, it is a program that identifies sites where hazardous substances have been or might have been released into the environment to ensure that they are cleaned up.

conditionally exempt small-quantity generator (CESQG)—those that generate no more than 100 kilograms of hazardous waste per month. Other than the hazardous waste determination requirement in 40 CFR 262.11, CESQGs are exempt from RCRA provided that they do not exceed certain quantity limits for hazardous waste storage or generation.

container—any portable device in which a material is stored, transported, treated, disposed of, or otherwise handled.

contingency plan—a document setting out an organized, planned, and coordinated course of action to be followed in case of a fire, explosion, or release of hazardous waste constituents that could threaten human health or the environment.

co-product—a material that is not a primary product, but is an incidentally produced product, which can be used in lieu of a commercially available product.

corrosive—a material that causes visible destruction of, or irreversible alterations in living tissue by chemical action at the site of contact.

designated facility—an authorized hazardous waste treatment, storage, or disposal facility that has been designated on a manifest by a generator to receive shipment of hazardous waste.

disposal—the discharge, deposit, injection, dumping, spilling, leaking, or placing of any solid waste or hazardous waste into or on any land or water so that such solid waste or hazardous waste or any constituent thereof may enter the environment or be emitted into the air or discharged into any waters, including groundwater.

end of pipe—residue from a process after pollution control equipment, such as sludge removed from solvent recovery or scrubber solution from a process scrubber.

EPA hazardous waste number—a number assigned by EPA to each hazardous waste listed in 40 CFR 261 Subpart D from specific and nonspecific sources that EPA has identified (F, K, P, U) and to each characteristic identified in 40 CFR 261 Subpart C, which includes wastes with ignitable (D001), corrosive (D002), reactive (D003), and toxic (D004–D043) characteristics.

EPA identification number—a number assigned by the EPA to each hazardous waste generator; transporter; and treatment, storage, or disposal facility. Identification numbers are facility specific, except for the transporter, which has one number for all operations.

flash point—the minimum temperature at which liquid gives off vapor in sufficient concentration to ignite when tested in three types of laboratory methods: Tagliabue Cleveland; Pensky-Martens Closed; or Setaflash Closed testers.

free liquids—liquids that readily separate from the solid portion of a waste under ambient temperature and pressure.

generator—any person whose process produces a hazardous waste in excess of 100 kilograms per month or acutely hazardous waste in excess of 1 kilogram per month, or whose actions first cause a hazardous waste to become subject to regulation.

grounding—the procedure used to carry an electrical charge to ground through a conductive path. A typical ground may be connected directly to a conductive pipe or a grounding bus and ground rod.

groundwater—water below the land surface in a zone of saturation.

HSWA—The Hazardous and Solid Waste Amendments of 1984 expanded RCRA and significantly increased regulatory control over hazardous waste handling and disposal. As a result of HSWA, disposal practices for certain wastes were banned, substances were added as listed wastes, minimum technology standards were specified, and the EPA's enforcement powers were expanded.

hazardous waste—includes those solid wastes that (1) have not been excluded from the definition of a hazardous waste, (2) have been listed as hazardous wastes by the EPA, (3) exhibit one or more of the characteristics of hazardous waste, or (4) have been mixed with a hazardous waste.

HAZWOPER—29 CFR 1910.120; the term used to describe the OSHA Hazardous Waste Operations and Emergency Response regulation. Regulatory guidelines are established for the protection of workers cleaning up a Superfund remediation site under CERCLA and hazardous material/waste spills in industrial, commercial, and public emergency response scenarios.

incompatible waste—a hazardous waste that is unsuitable for (1) placement in a particular device or facility because it may cause corrosion or decay of contaminated materials; or (2) commingling with another waste or material under controlled conditions because the commingling might produce heat or pressure, fire or explosion, violent reaction, toxic dusts, mists, fumes or gases, or flammable fumes or gases.

infectious waste—(1) equipment, instruments, utensils, and fomites (substances that may harbor or transmit pathogenic organisms) of a disposable nature from the rooms of patients who are suspected to have or have been diagnosed with a communicable disease; (2) laboratory wastes, such as pathological specimens; (3) surgical operating room pathologic specimens and similar disposable materials from outpatient areas and emergency rooms.

lab pack—the process of placing smaller containers such as glass jars inside an open head drum for proper shipment or disposal.

land disposal restrictions (LDR)—also known as the *land ban*. These restrictions under RCRA prohibit any land disposal of untreated hazardous wastes. LDRs establish treatment standards that must be met prior to any land application of hazardous wastes, and they recommend specific treatment methods for meeting standards.

large-quantity generator (LQG)—a generator that generates more than 1000 kilograms of hazardous waste in a calendar month.

large-quantity handler of universal waste (LQHUW)—a universal waste handler that accumulates 5000 kilograms or more of total universal waste (batteries, pesticides, or thermostats, calculated collectively) at any time. This designation as an LQH of universal waste is retained through the end of the calendar year in which 5000 kilograms or more of total universal waste is accumulated.

listed wastes—To make it simple to determine whether a waste is hazardous, the EPA has evaluated some wastes and determined them to be hazardous by applying their own criteria. If the waste is "listed," no sampling, analysis, or detailed knowledge development is required to determine if it is hazardous.

manifest—the cradle-to-grave paperwork that must accompany a shipment of hazardous waste as it moves from the generator to the transporter and eventually to the TSDF. The generator originates and signs the manifest (EPA form 8700-22 and, if necessary, form 8700-22A) in accordance with the instructions included in Appendix to 40 CFR Part 262.

mixed waste—waste that contains both hazardous and nonhazardous source (listed or characteristic) constituents.

point of waste origination—(1) When the facility owner or operator is the generator of hazardous waste, point of waste origination is the point at which a solid waste produced by a system, process, or waste management unit is determined to be hazardous waste as defined in 40 CFR part 261. (2) When the facility owner and operator are not the generator of hazardous waste, point of waste origination means the point at which the owner or operator accepts delivery or takes possession of the hazardous waste (40 CFR 265.1081).

RCRA—the Solid Waste Disposal Act as amended by the Resource Conservation and Recovery Act of 1976. It is the federal statute that regulates the generation, treatment, storage, disposal recycling, and/or transportation of solid and hazardous waste.

recycling—using, reusing, or reclaiming materials/waste, including processes that regenerate a material or recover a usable product from it.

small-quantity generator (SQG)—a regulated facility that generates more than 100 kilograms and less than 1000 kilograms of hazardous waste in a calendar month.

small-quantity handler of universal waste (SQHUW)—a universal waste handler, as defined in 40 CFR 273.6, which does not accumulate more than 5000 kilograms total of universal waste (batteries, pesticides, or thermostats, calculated collectively) at any time.

solid waste—any garbage, refuse, sludge, or other waste materials not excluded by definition. Exclusions (materials not defined as "solid wastes") include domestic sewage and any mixture of other wastes that passes through a sewage system to a publicly owned treatment works (POTW); industrial wastewater discharges that are point source discharges subject to regulation under the Clean Water Act; irrigation return flows; and in situ mining materials. The EPA defines hazardous waste as a subset of solid waste.

source reduction—any practice that (1) reduces the amount of any hazardous substance, pollutant, or contaminant entering any waste stream or otherwise released into the environment (including fugitive emissions) prior to recycling, treatment, and disposal; (2) reduces the hazards to public health and the environment associated with the release of such substances, pollutants, or contaminants.

storage—the holding of hazardous waste for a temporary period; at the end the hazardous waste is treated, disposed of, or stored elsewhere. Facilities are required to have an RCRA permit for storage of hazardous waste for more than 90 days; storage for less than 90 days does not require an RCRA permit.

surface impoundment—a facility or part of a facility that is a natural topographic depression, human-made excavation, or diked area formed primarily of earthen materials (it can be lined with human-made materials), that is designed to hold an accumulation of liquid wastes containing free liquids, and that is not an injection well. Examples of surface impoundments are holding, storage, settling, and aeration pits, ponds, and lagoons.

tank—a stationary device designed to contain an accumulation of hazardous waste that is constructed primarily of nonearthen materials (e.g., wood, concrete, steel, plastic).

toxicity characteristic leaching procedure (TCLP)—a laboratory procedure whereby a waste is immersed in a standardized "garbage juice" (organic acid buffer solution) to determine whether any constituents of concern are leached out in amounts sufficient to pose a hazard (40 CFR 261.24 and EPA publication "SW-846").

TSCA—the Toxic Substance Control Act—a federal environmental legislation administered by the EPA. TSCA regulates the manufacture, handling, and use of materials classified as toxic substances. Asbestos abatement (including disposal) falls under the jurisdiction of the TSCA regulation.

transporter—a person transporting hazardous waste that requires a manifest within the United States. Those who move hazardous waste on-site are not considered transporters. Transporters must comply with 40 CFR Part 263.

treatment—any method, technique, or process, including neutralization, designed to change the physical, chemical, or biological character or composition of any hazardous waste so as to neutralize such waste, or so as to recover energy or material resources from the waste, or so as to render the waste nonhazardous or less hazardous; safer to transport, store, or dispose of; or amenable for recovery, amenable for storage, or reduced in volume.

TSDF—a permitted facility that treats, stores, and/or disposes of hazardous wastes.

underground storage tanks (USTs)—any one or a combination of tanks (including underground connective pipes) that is used to accumulate regulated substances the volume of which (including the volume of the underground connective pipes) is 10 percent or more beneath the surface of the ground. Regulated substances include hazardous chemical products regulated under CERCLA and petroleum products. USTs containing hazardous waste are regulated under Subtitle C of RCRA.

waste determination—a methodology to identify the composition of a material by means of testing or by employing a generator's knowledge of the process.

waste minimization—the use of materials, processes, or practices that reduce or eliminate the creation of pollutants or wastes at the source. It includes practices that reduce the use of hazardous materials, energy, water, or other resources, and practices that protect natural resources through conservation or more efficient use.

References

American Chemical Council (ACC)/Synthetic Organic Manufacturers Association (SOCMA). 1996. *Compliance Assistance Tool for RCRA Subpart CC, 40 CFR 264 and 265*. Washington, D.C.: ACC/SOCMA.

American Chemical Council (ACC). 1991. *Pollution Prevention Resource Manual*. Washington, D.C.: ACC. Aspen Editorial Staff. 2008. *RCRA Regulation and Keyword Index*. New York: Aspen Publishers.

Conversion Systems Incorporated. 1993. *A Glossary of Environmental Terms, Expressions, and Acronyms*. Horsham, Pennsylvania: Conversion Systems, Inc.

Keegan, Robert. 2007–2008. *Hazardous Materials, Substances, and Waste Compliance Guide*. Kutztown, Pennsylvania: Hazardous Materials Publishing Company, Inc.

Little, Arthur D. 1995. *Environmental Regulatory Field Guide*. Boston, Massachusetts: Arthur D. Little.

Stewart R. Browne Manufacturing Co., Inc. 2007. *Grounding and Bonding Applications for Control of Static Electricity*. Atlanta, Georgia: Stewart R. Browne Manufacturing Co., Inc.

USEPA. 1994. *Waste Analysis at Facilities That Generate, Treat, Store, and Dispose of Hazardous Wastes:—A Guidance Manual*. Washington, D.C.: Office of Solid Waste and Emergency Response.

______. 2001. *Land Disposal Restrictions, Summary of Requirements*. Washington, D.C.: Office of Solid Waste and Emergency Response.

______. 2006. *RCRA Orientation Manual, 2006 Edition,*. Washington, D.C.: Office of Solid Waste and Emergency Response.

United States Department of Transportation (USDOT). 2008. *North American Emergency Response Guidebook*. Washington, D.C.: Office of Hazardous Materials Initiatives and Training, Research, and Special Programs Administration.

Zoch, Robert. 1986. *RCRA Handbook*. Concord, Massachusetts: ERT Engineering.

Index

I

J

K

L

M

N

O

P

R

S

T

U

V

W

X

PGIL2024USA